T0275932

SpringerBriefs in Electrical and Computer Engineering

Signal Processing

Series editors

Woon-Seng Gan, Singapore, Singapore
C.-C. Jay Kuo, Los Angeles, USA
Thomas Fang Zheng, Beijing, China
Mauro Barni, Siena, Italy

More information about this series at http://www.springer.com/series/11560

Samuel Davey · Neil Gordon
Ian Holland · Mark Rutten · Jason Williams

Bayesian Methods
in the Search for MH370

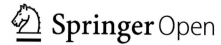

Samuel Davey
National Security and ISR Division
Defence Science and Technology Group
Edinburgh, SA
Australia

Mark Rutten
National Security and ISR Division
Defence Science and Technology Group
Edinburgh, SA
Australia

Neil Gordon
National Security and ISR Division
Defence Science and Technology Group
Edinburgh, SA
Australia

Jason Williams
National Security and ISR Division
Defence Science and Technology Group
Edinburgh, SA
Australia

Ian Holland
Cyber and Electronic Warfare Division
Defence Science and Technology Group
Edinburgh, SA
Australia

ISSN 2191-8112   ISSN 2191-8120   (electronic)
SpringerBriefs in Electrical and Computer Engineering
ISSN 2196-4076   ISSN 2196-4084   (electronic)
SpringerBriefs in Signal Processing
ISBN 978-981-10-0378-3   ISBN 978-981-10-0379-0   (eBook)
DOI 10.1007/978-981-10-0379-0

Library of Congress Control Number: 2015960765

© Commonwealth of Australia 2016. This book is published open access.
Open Access This book is distributed under the terms of the Creative Commons Attribution-NonCommercial 4.0 International License (http://creativecommons.org/licenses/by-nc/4.0/), which permits any noncommercial use, duplication, adaptation, distribution and reproduction in any medium or format, as long as you give appropriate credit to the original author(s) and the source, a link is provided to the Creative Commons license and any changes made are indicated.
The images or other third party material in this book are included in the work's Creative Commons license, unless indicated otherwise in the credit line; if such material is not included in the work's Creative Commons license and the respective action is not permitted by statutory regulation, users will need to obtain permission from the license holder to duplicate, adapt or reproduce the material.
This work is subject to copyright. All commercial rights are reserved by the Publisher, whether the whole or part of the material is concerned, specifically the rights of translation, reprinting, reuse of illustrations, recitation, broadcasting, reproduction on microfilms or in any other physical way, and transmission or information storage and retrieval, electronic adaptation, computer software, or by similar or dissimilar methodology now known or hereafter developed.
The use of general descriptive names, registered names, trademarks, service marks, etc. in this publication does not imply, even in the absence of a specific statement, that such names are exempt from the relevant protective laws and regulations and therefore free for general use.
The publisher, the authors and the editors are safe to assume that the advice and information in this book are believed to be true and accurate at the date of publication. Neither the publisher nor the authors or the editors give a warranty, express or implied, with respect to the material contained herein or for any errors or omissions that may have been made.

Printed on acid-free paper

This Springer imprint is published by SpringerNature
The registered company is Springer Science+Business Media Singapore Pte Ltd.

Foreword I

Uncertainty is all pervasive—whether it relates to everyday personal choices and actions, or as background to business and policy decisions, or economic and climate predictions. In recent times, few things have attracted as much attention as the uncertainty surrounding the final whereabouts of MH370.

How to deal scientifically with uncertainty? Put simply, on the one hand there are events or outcomes of interest that we don't know; on the other hand, pieces of information that we judge relevant in some sense that we do know. We need to assess what we believe about the unknowns, given the knowns.

Formalising our measure of uncertainty in terms of probabilities, the scientific approach is encapsulated in the so-called Bayesian statistical paradigm, in which beliefs about the unknowns are quantified by a probability measure conditional on what we know.

But typically, our state of knowledge itself gets modified over time and a method is therefore needed to refine and update beliefs as new information is acquired and assembled. The logical, mathematical rule for carrying out this updating is Bayes theorem, hence the term Bayesian Methods to describe the analytic and computational toolkit that has been developed for updating beliefs as evidence changes or is added to.

It is this toolkit that has been employed in the search for MH370 and this fascinating book provides a blow-by-blow case account of how the various strands of evidence have been brought together to give an overall probabilistic assessment of the final whereabouts of the plane.

This has been an extremely complex task and the authors are to be congratulated on setting out systematically and coherently the science and mathematics driving the evidential equations. But, in addition to the complex modelling there remains the task of pulling out the Bayesian probability messages from the tangle of data that has been assembled. The computational methods for achieving this are of relatively recent origin and, on a personal note, I am delighted to have played a small part, with Dr. Neil Gordon, in the signal processing revolution that is now the particle filter method of analysis.

Dr. Gordon and his colleagues have employed these methods to provide a marvellous case study demonstrating the power of mathematical modelling and computation to attack one of the most intractable uncertain puzzles of recent times.

London, UK   Professor Sir Adrian Smith
November 2015   FRS, Vice Chancellor
                                                                         University of London

Foreword II

The disappearance of the Malaysian Airlines flight MH370 from air traffic control radar on the evening of 8 March 2014 with 227 passengers and 12 crew on board was a tragedy and remains a mystery until we are able to locate the flight recorder or the wreckage. Only then can we unravel what actually occurred.

Initially, the multinational search and recovery effort was focused on the Gulf of Thailand and the South China Sea. However, following closer analysis of Inmarsat data and flight path projections the search shifted to the Indian Ocean.

The Malaysian government accepted the Australian government's offer to take the lead in the search and recovery operation in the southern Indian Ocean. Australia, led by the Australian Transport Safety Bureau (ATSB), took responsibility for defining the underwater search area in 28 April 2014.

The Defence Science and Technology (DST) Group, being the only Australian government agency with the combined knowledge and experience in the scientific disciplines to support the search for MH370, became involved from May 2014. Defence scientists contributed a range of expertise across a spectrum of technologies to the search: underwater acoustics, satellite communication systems and statistical data processing.

DST Group's position as a trusted government adviser with stewardship of the full range of defence technologies has been critical to our ability to contribute to the multi-agency—and multinational-search. Our ability to provide high quality, internationally respected, rapid response is built on our deep foundational research capability. Our support to the MH370 incident is an example of scientific research feeding directly into an active operational search.

The quest for the MH370 demonstrates our ability to 'work the full problem' with a team of internationally-recognised experts contributing to the ATSB working group. Indeed, DST Group is proud to have been able to contribute its world-renowned expertise to the ATSB-led search. However, this book is not about the search for MH370. Rather, it focuses on the work to define the search zone.

We are particularly fortunate to have Dr. Neil Gordon leading the DST Group team. Dr. Gordon is recognised internationally as an expert in statistical data processing and, in particular, the dynamic Bayesian estimation methods deployed in this activity.

Working alongside Dr. Gordon are other Defence scientists—Dr. Samuel Davey, Dr. Mark Rutten and Dr. Jason Williams who are all experts in target tracking and multi-sensor fusion, and Dr. Ian Holland who specialises in satellite and wireless communications.

DST Group's expertise and unique capabilities will continue to contribute to the ongoing search for the MH370, led by the ATSB and in collaboration with other Australian and international agencies.

Canberra, Australia   Dr. Alex Zelinsky
November 2015   Chief Defence Scientist
                                                        Defence Science and Technology
                                                              Department of Defence

Acknowledgements

This book is a summary of our work as members of the MH370 Flight Path Reconstruction group. The group comprised members from Inmarsat, Thales, Boeing, US National Transportation Safety Board (NTSB) and the UK Air Accidents Investigation Branch (AAIB). The group has been expertly led by the Australian Transport Safety Bureau (ATSB).

To all our friends and colleagues at the ATSB: During many intense and stressful times you always remained calm and assured. You are dedicated professionals.

To the Flight Path reconstruction working group: We thank you all for freely sharing your expert knowledge and for working together both independently and collaboratively in such a positive way. We also thank you for sharing your forthright opinions and views and questioning everything.

To our friends and colleagues at the Defence Science and Technology Group: We would particularly like to thank Gerald Bolding and Balachander Ramamurthy for their support to the satellite data analysis and David Liebing for all his efforts in drawing the team together.

The material in Chap. 11 related to the Reunion Island debris find was developed with David Griffin from CSIRO Marine and Atmospheric Research in Hobart. Thank you David for hosting us and sharing your knowledge.

We thank Sanjeev Arulampalam (DST Group, Australia), Vaughan Clarkson (University of Queensland, Australia), Simon Godsill (Cambridge University, UK), Fredrik Gustafsson (Linköping University, Sweden), Simon Maskell (Liverpool University, UK), and Thomas Schön (Uppsala University, Sweden) for their thoughtful and helpful comments and suggestions. We thank Brian Anderson for communicating some minor corrections to the presentation.

All those involved in the search for MH370 remain totally committed to finding the aircraft and helping find closure for the families involved.

Contents

About the Authors

Samuel Davey received the Bachelor of Engineering, Master of Mathematical Science and Ph.D. degrees from the University of Adelaide, Australia, in 1996, 1999 and 2003, respectively. Since 1995 he has worked for the Defence Science and Technology Group, Australia, in the areas of target tracking, tracker performance assessment and multi-sensor fusion. He is a Visiting Research Fellow at the University of Adelaide and a Senior Member of the IEEE.

Neil Gordon received a Ph.D. in Statistics from Imperial College London in 1993. He was with the Defence Evaluation and Research Agency in the UK until 2002 working on missile guidance and statistical data processing. He is best known for initiating the particle filter approach to nonlinear, non-Gaussian dynamic estimation, which is now in widespread use throughout the world in many diverse disciplines. He is the co-author/co-editor of two books on particle filtering. In 2002 he moved to the Defence Science and Technology Group in Adelaide, Australia, where he is currently head of Data and Information Fusion. In 2014 he became an Honorary Professor with the School of Information Technology and Electrical Engineering at the University of Queensland. He is a Senior Member of the IEEE.

Ian Holland received the Bachelor of Electronic and Communication Engineering in 2000 and a Ph.D. in wireless telecommunications in 2005, both from Curtin University of Technology, Western Australia. Since then he has held positions in the University of Western Australia, the Institute for Telecommunications Research at the University of South Australia, EMS Satcom Pacific and Lockheed Martin Australia. Since January 2011, Ian has been working as a Research Scientist in Protected Satellite Communications at the Defence Science and Technology Group.

Mark Rutten received the Bachelor of Science, Bachelor of Engineering and Master of Mathematical Science from the University of Adelaide in 1995, 1996 and 1999, respectively, and a Ph.D. from the University of Melbourne in 2005 on Multipath Tracking for Over the Horizon Radars. He has worked on data and information fusion for the Defence Science and Technology Group since 1996. His research interests include nonlinear state estimation, target tracking and multi-sensor fusion.

Jason Williams received degrees of Bachelor of Engineering in Electronics and Bachelor of Information Technology from Queensland University of Technology in 1999, Master of Science in Electrical Engineering from the United States Air Force Institute of Technology in 2003, and Ph.D. in Electrical Engineering and Computer Science from Massachusetts Institute of Technology in 2007. He worked for several years as an engineering officer in the Royal Australian Air Force, before joining Australia's Defence Science and Technology Group in 2007. He is also an Adjunct Senior Lecturer at the University of Adelaide. His research interests include target tracking, sensor resource management, Markov random fields and convex optimisation.

Acronyms

9M-MRO	Registration number of the accident aircraft
AAIB	Air Accidents Investigation Branch (United Kingdom)
ACARS	Aircraft Communications Addressing and Reporting System
ACCESS-G	Australian Community Climate and Earth-System Simulator Global model
AES	Aircraft Earth Station (the aircraft satellite communications unit)
ATC	Air Traffic Control
ATSB	Australian Transport Safety Bureau
BFO	Burst Frequency Offset
BTO	Burst Time Offset
cdf	Cumulative Distribution Function
CI	Cost Index (used to automatically control air speed)
CMH	Constant Magnetic Heading navigation
CMT	Constant Magnetic Track navigation
CSIRO	Commonwealth Scientific and Industrial Research Organisation
CTH	Constant True Heading navigation
CTT	Constant True Track navigation
DST Group	Defence Science and Technology Group, Australia
GES	Ground Earth Station (the ground component of the satellite communications system)
HPD	Highest Posterior Density
JACC	Joint Agency Coordination Centre
KL	Kuala Lumpur
LNAV	Lateral NAVigation
NEES	Normalised Estimation Error Squared
NTSB	National Transportation Safety Board (United States)
OU	Ornstein–Uhlenbeck process

pdf	Probability Density Function
RMS	Root Mean Squared
SATCOM	Satellite communications
SIR	Sample Importance Resample particle filter
UTC	Coordinated Universal Time

Chapter 1
Introduction

On 7 March 2014 at 16:42,[1] Malaysian Airlines flight MH370 departed from Kuala Lumpur (KL) International Airport bound for Beijing. There was a total of 239 persons on board (227 passengers and 12 crew). The aircraft was a Boeing 777-200ER registered as 9M-MRO. The aircraft lost contact with Air Traffic Control (ATC) during a transition between Malaysian and Vietnamese airspace. The last recorded radio transmission from MH370 was at 17:19. Over the following days, an intensive air and sea rescue search was made around the last reported position of the aircraft in the Gulf of Thailand without success. It then became clear that satellite communication messages between the aircraft and one member of Inmarsat's constellation of geosynchronous satellites were crucial to defining the search zone for the aircraft. Satellite communication systems involve transmissions between multiple terminals using a satellite. The aircraft 9M-MRO communicated with the Inmarsat ground station in Perth, Western Australia, via the Indian Ocean Region satellite Inmarsat-3F1. The data available for MH370 is mostly comprised of approximately hourly "handshake" transmissions initiated by the ground station for aircraft that have not communicated in the preceding hour. No explicit information relating to the aircraft terminal location is contained in the messages; however, the messages contain metadata which can be processed to produce estimates of the flight path and final location.

Inmarsat conducted a rapid and innovative analysis of the data which placed the aircraft in the Australian search and rescue zone on an arc in the Southern part of the Indian Ocean. On 17 March 2014, the Australian Maritime Safety Authority took responsibility of the search and rescue operation. Subsequently the Joint Agency Coordination Centre (JACC) was established on 30 March 2014 to coordinate the Australian Government's support for the search and over the following weeks an intensive aerial and surface search was conducted by an international team.

---

[1]All times are given in Coordinated Universal Time (UTC) in format hh:mm:ss. Local time in Malaysia and Western Australia is 8 h ahead of UTC and local time in West Indonesia is 7 h ahead of UTC.

© Commonwealth of Australia 2016
S. Davey et al., *Bayesian Methods in the Search for MH370*, SpringerBriefs in Electrical and Computer Engineering, DOI 10.1007/978-981-10-0379-0_1

On 28 April 2014, the aerial search concluded and the search moved to an underwater phase. The Australian Transport Safety Bureau (ATSB) took responsibility for defining the underwater search area. The ATSB convened a flight path prediction working group in order to bring together experts in satellite communication systems and statistical data processing and apply novel data analysis techniques to estimate the most likely final location of MH370. This working group consisted of representatives from the following organisations: Air Accidents Investigation Branch (UK); Boeing (US); Inmarsat (UK); National Transportation Safety Board (US); Thales (UK); and the Defence Science and Technology (DST) Group (Australia). The working group developed new methods to analyse the Inmarsat data and validated those methods. The ATSB released a report summarising the findings of the working group in August 2014 [3] and have subsequently released updates in October 2014 [4] and December 2015 [5].

In this book we detail the statistical approach adopted by the DST Group team to analyse the available data and produce a probability density function (pdf) of the accident aircraft's final location. In Chap. 2, we start by detailing a summary of the events and listing of the available data. In Chap. 3, we describe the Bayesian framework, upon which our method is formulated. This approach requires several ingredients. The first of these is a prior distribution; this is built on the primary radar data described in Chap. 4. The second ingredient consists of a set of likelihood functions detailing how the available measurements are linked to the aircraft state; the measurement models and error statistics are characterised in Chap. 5. The final ingredient is a stochastic model describing the possible dynamic trajectories of the aircraft. Chapters 6 and 7 describe our models for aircraft dynamics during cruise and manoeuvre respectively. The resulting set of (prior, likelihood, dynamics) enables us to calculate the probability distribution of aircraft trajectories. However since the models are nonlinear and non-Gaussian we are required to use numerical methods for the calculation; our particle filtering approach is described in Chap. 8. The Bayesian method that was developed has been validated against a number of earlier flights of the accident aircraft 9M-MRO, where accurate measurements of the aircraft location were available from the aircraft's logging system. These results are detailed in Chap. 9. The method has also been applied to the data available for the accident flight; the resulting probability distribution, which defines the search zone, is described in Chap. 10. The search zone is defined by combining our pdf from the analysis of the satellite data with a kernel describing the distribution of aircraft motion during descent, which was defined by expert accident investigators from the ATSB. Thus, any adjustment to the assumptions made about the descent (and hence the kernel describing its distribution) yields a change to the search region. Finally, in Chap. 11, we discuss on-going work including the impact of the flaperon wreckage discovery on Reunion Island.

Being statistical in nature, this book seeks to identify aircraft paths that are most likely, given models which incorporate sensor measurements, commonly observed commercial aircraft motion, and air transport safety investigator and manufacturer expert advice and assessments. The goal is not to identify the complete and exhaustive set of all possible aircraft paths, but rather the subset of those trajectories that are most probable. The area covered by the set of end-points of all possible paths is prohibitively large and the overwhelming majority of this area contains vanishingly small probability. In determining how to allocate finite search resources, priority is directed to the area containing the highest probability trajectories. The proposed method incorporates a rich variety of possible paths, and calculates a probability distribution based on the well-established, rigorous Bayesian toolkit, which automatically trades the complexity of the model against the match to the observed measurements.

The ATSB maintains a website http://www.atsb.gov.au/mh370.aspx with comprehensive information about the ongoing search and we refer readers to this resource for the latest news, data, maps, videos, reports and operational updates. Any comments or information related to the search or the analysis in this book should be sent to atsbinfo@atsb.gov.au. Feedback related to the analysis in this book can also be sent to MH370@dsto.defence.gov.au.

1.1 Summary of Results

The key outcome of this book is a probability distribution of the final aircraft location based on a Bayesian analysis, using models constructed through detailed study of the measurement noise statistics and commercial aircraft motion, incorporating assessments of likely operating parameters from expert accident investigators. The distribution is shown in Fig. 1.1, illustrated as a contour plot, where the colour

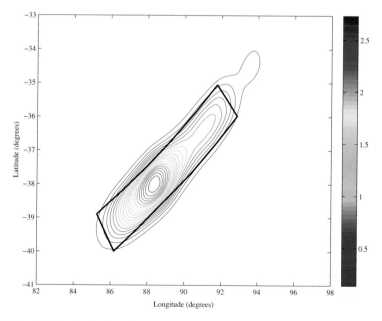

Fig. 1.1 Probability density function of final location of MH370. Indicative search area (as of November 2015) marked with *solid black line*

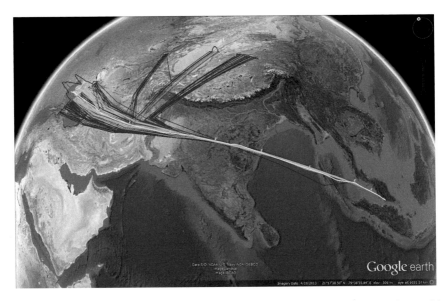

Fig. 1.2 Result of validation analysis applied to the 9M-MRO flight from KL to Amsterdam on 26 February 2014

indicates the likelihood, with red being most likely, and blue being least likely. The black rectangle shows the indicative search area as at November 2015. Full details can be found in Chap. 10. In order to orient the scale of the search being undertaken, the area of the indicative search region shown in Fig. 1.1 is 100,000 km^2. In contrast, the bounding region considered in the search for Air France flight 447[2] was a disc with radius 40 nm [40], corresponding to an area of 17,000 km^2.

The procedure used to generate the result in Fig. 1.1 was validated by applying the identical process to other flights for which the location was known. The flights examined included four flights for the same aircraft (9M-MRO) involved in the accident, as well as two others that were in flight at the same time as the accident. The communications logs for these flights were down-sampled to emulate the information available on the accident flight. An example of the result is shown in Fig. 1.2. The white line shows the true path taken by the aircraft, while other coloured lines show potential trajectories, coloured by their probability, again with red indicating the most likely paths and blue indicating the least likely. The probability distribution at the time of the final measurement is shown as a red line, with the likelihood encoded through the height above the earth.

Ten different subsets of measurements were used from each of the six flights to make a total of 60 validation experiments. In all cases, the true final aircraft

---

[2]Air France Flight 447 disappeared over the South Atlantic Ocean on 1 June 2009. The initial underwater search consisted of three phases over approximately one year. After the third phase of the search, US company Metron conducted a Bayesian analysis of the available data. Debris was found in April 2011 close to the final reported location of the aircraft.

location was contained within the 85 % probability region. This indicates that the probability distribution produced by the method is slightly conservative, otherwise only around 50 of the 60 experiments should be in the 85 % region. This is because the dynamics model allows for a wider range of aircraft manoeuvres than are actually experienced by typical commercial flights. Such a model is appropriate because the accident flight was not a typical commercial flight. Full details of the validation experiments can be found in Chap. 9.

Open Access This chapter is distributed under the terms of the Creative Commons Attribution-NonCommercial 4.0 International License (http://creativecommons.org/licenses/by-nc/4.0/), which permits any noncommercial use, duplication, adaptation, distribution and reproduction in any medium or format, as long as you give appropriate credit to the original author(s) and the source, a link is provided to the Creative Commons license and any changes made are indicated.

The images or other third party material in this chapter are included in the work's Creative Commons license, unless indicated otherwise in the credit line; if such material is not included in the work's Creative Commons license and the respective action is not permitted by statutory regulation, users will need to obtain permission from the license holder to duplicate, adapt or reproduce the material.

Chapter 2
Factual Description of Accident and Available Information

The detailed chronological factual statement of known information about flight MH370 is given at [35]. A brief summary is given here sufficient to put the analysis in the rest of the book in context.

On 7 March 2014 at 16:42, flight MH370 departed from KL bound for Beijing; this is marked as event 1 in Fig. 2.1. Initially, everything about the flight proceeded as normal. The Mode S transponder system on-board the aircraft was responding as expected to interrogation from the ATC Secondary Surveillance Radar up to the time when it was lost on the ATC radar screen at 17:21:13, marked as event 4. No message was received from the aircraft to report a system failure. Similarly the on-board Aircraft Communications Addressing and Reporting System (ACARS) reported as expected at 17:07:29, this event is marked as 2 in Fig. 2.1. This message contained a collation of six reports generated at five minute intervals by the system from 16:41:43 until 17:06:43. These reports contained information about the aircraft position and motion such as latitude, longitude, altitude, air temperature, air speed, wind direction, wind speed, and true heading. The ACARS position reports are scheduled to be transmitted at thirty minute intervals during cruise. The next scheduled report at 17:37 was not received. The last recorded radio transmission with the crew of MH370 occurred at 17:19:30 as the aircraft was instructed to contact Vietnamese ATC on leaving the Malaysian Flight Information Region. This communication is marked as event 3. At 17:39:06 Vietnamese ATC contacted KL ATC to query the whereabouts of MH370, who initiated the efforts of several countries' ATC to establish the location, without success. Malaysian military radars were subsequently able to show primary radar returns associated with MH370 deviating from the flight plan almost immediately after the loss of Secondary Surveillance Radar at 17:21:13 by making a left turn to end up travelling in a South Westerly direction. Radar returns show the aircraft travelling back across Malaysia before turning near Penang Island and travelling in a North Westerly direction up the Straits of Malacca. The final primary radar return was recorded at 18:22:12, which is marked as event 5 in Fig. 2.1.

© Commonwealth of Australia 2016
S. Davey et al., *Bayesian Methods in the Search for MH370*, SpringerBriefs
in Electrical and Computer Engineering, DOI 10.1007/978-981-10-0379-0_2

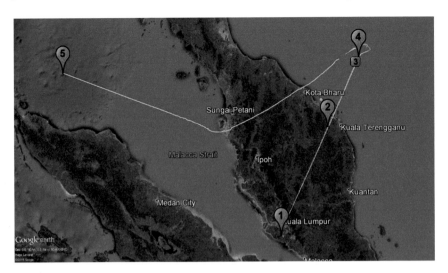

Fig. 2.1 Chronological sequence of events in MH370 disappearance. *1* = Take off; *2* = Final ACARS transmission; *3* = Final radio contact; *4* = Final Mode S transmission; *5* = Final primary radar detection

The ACARS system also provides message communication between the aircraft and its ground base. At 18:03:23 Malaysian Airlines Operations Centre sent a message asking the crew to contact Vietnamese ATC immediately. The downlink message showed that this failed to reach the aircraft. The message was auto transmitted every two minutes until 18:43:33 with no success [35]. The next communication on the SATCOM system was a log-on request from the aircraft at 18:25. This communication contains no location information from the aircraft but it does however contain timing and frequency metadata which turn out to be crucial in estimating where MH370 went after the final radar return at 18:22:12. The timing measurement is called the Burst Timing Offset (BTO) and the frequency measurement the Burst Frequency Offset (BFO). At 18:39:52 there was an unanswered attempt to call the aircraft from ATC on the ground to air telephone link. The failed communication data log contains frequency but not timing metadata.

If there has been no SATCOM activity for sixty minutes then the ground station automatically initiates a handshake confirming presence of the aircraft. If the aircraft receives this query then an automatic response is sent indicating that it is still logged on to the SATCOM network. These handshake responses contain both timing and frequency metadata. Handshake messages occurred at 19:41:00, 20:41:02, 21:41:24 and 22:41:19. At 23:15:58 there was a second unanswered attempt at a satellite telephone call, again giving frequency metadata. There was a ground station initiated handshake at 00:10:58 (timing and frequency metadata). Finally, at 00:19:29 the aircraft SATCOM system initiated another log-on request and this was the last SATCOM transmission received from the aircraft. Due to its timing, this log-on request is believed to correspond to fuel exhaustion and subsequent

Table 2.1 Summary of SATCOM data available for MH370

Event	Time (UTC)	BTO	BFO
Aircraft departed KL	16:42	Y	Y
Last ACARS transmission	17:07	Y	Y
AES initiated log on	18:25	Y	N*
AES access request	18:28	Y	Y
Unanswered ground to air telephone call	18:39	N	Y
GES initiated handshake	19:41	Y	Y
GES initiated handshake	20:41	Y	Y
GES initiated handshake	21:41	Y	Y
GES initiated handshake	22:41	Y	Y
Unanswered ground to air telephone call	23:15	N	Y
GES initiated handshake	00:10	Y	Y
AES initiated logon	00:19	Y	N*

Measurements marked with an *asterisk* are available but cannot be used as discussed in the text

activation of the auxiliary power unit [5]. There was no response to the ground station initiated handshake request at 01:15:56. The log-on messages at 18:25 and 00:19 contain BFO measurements but the equipment was not in steady state. The values of the BFO measurements were deemed to be unreliable at these times and cannot be used.

The key pieces of information available to us to estimate the MH370 flight path are the final radar detection at 18:22:12 and the timing and frequency metadata associated with the infrequent SATCOM messaging that subsequently occurred up until 00:19. Table 2.1 summarises the available measurement data. The messages at 18:25 and 00:19 occurred during transient phases of operation for the SATCOM equipment so the BFO values reported for these times cannot be used. In the next two chapters we describe the Bayesian approach, and then how the satellite communication system works and how the timing and frequency metadata can be related to aircraft location, allowing us to build a likelihood function for our Bayesian analysis.

Open Access This chapter is distributed under the terms of the Creative Commons Attribution-NonCommercial 4.0 International License (http://creativecommons.org/licenses/by-nc/4.0/), which permits any noncommercial use, duplication, adaptation, distribution and reproduction in any medium or format, as long as you give appropriate credit to the original author(s) and the source, a link is provided to the Creative Commons license and any changes made are indicated.

The images or other third party material in this chapter are included in the work's Creative Commons license, unless indicated otherwise in the credit line; if such material is not included in the work's Creative Commons license and the respective action is not permitted by statutory regulation, users will need to obtain permission from the license holder to duplicate, adapt or reproduce the material.

Chapter 3
The Bayesian Approach

Bayesian inference methods [9] provide a well-studied toolkit for calculating a distribution of a quantity of interest given observed evidence (measurements). As such, they are well-suited for calculating a probability distribution of the final location of the aircraft given the data available from the Inmarsat satellite communication system. The resulting probability distribution is essential to prioritise search efforts. In this chapter, we provide a brief introduction to Bayesian methods. We assume a reasonable background in probability theory; the interested reader is referred to excellent resources such as [8, 9, 10, 18, 36, 39, 43] if further detail is desired.

The required probability density function (pdf) is the probability of the aircraft location given the available data. Bayes' rule defines a method to calculate this pdf using prior information, including knowledge of how aircraft move, and a model of how the measured data relate to the aircraft location and velocity. Mathematically, Bayes' rule is

$$p(\mathbf{x}|\mathbf{z}) = \frac{p(\mathbf{x}, \mathbf{z})}{p(\mathbf{z})} \tag{3.1}$$

$$= \frac{p(\mathbf{z}|\mathbf{x})\, p(\mathbf{x})}{p(\mathbf{z})} \tag{3.2}$$

$$= \frac{p(\mathbf{z}|\mathbf{x})\, p(\mathbf{x})}{\int p(\mathbf{z}|\mathbf{x}')\, p(\mathbf{x}')\mathrm{d}\mathbf{x}'} \tag{3.3}$$

where the elements are:

1. $\mathbf{x}$ is the random variable, or the *state*, which is the quantity of interest (e.g., the position of the aircraft);
2. $\mathbf{z}$ is the measurement (e.g., the Inmarsat satellite communication data, which provides some form of positional data);
3. $p(\mathbf{x})$ is the prior pdf of the state (not incorporating the measurement, e.g., based on historical data);

© Commonwealth of Australia 2016
S. Davey et al., *Bayesian Methods in the Search for MH370*, SpringerBriefs in Electrical and Computer Engineering, DOI 10.1007/978-981-10-0379-0_3

4. $p(\mathbf{z}|\mathbf{x})$ is the pdf of the measurement conditioned on the state (e.g., this may be constructed by observing the distribution of measurements in cases where the state is known);
5. $p(\mathbf{x}|\mathbf{z})$ is the conditional pdf of interest (the posterior pdf), describing the distribution of state (e.g., aircraft location) taking into account the observed measurement.

The posterior probability density is based on the accumulated Inmarsat satellite communications data as well as all available contextual knowledge on the sensor characteristics, aircraft dynamic behaviour and environmental conditions and constraints. The method is based on the state space approach to time series modelling. Here, attention is focused on the state vector of a system. The state vector contains all relevant information required to describe the system under investigation at a given point in time. For example, in radar tracking problems this information would typically be related to the kinematic characteristics of the aircraft, such as position, altitude, speed, and heading. The measurement vector represents noisy observations that are related to the state vector. For example, the distance and bearing angle between the sensor and the object being measured. The state-space approach is convenient for handling multivariate data and nonlinear, non-Gaussian processes; it provides a significant advantage over traditional time series techniques for these problems; and has been extensively used in many diverse applications over the last 50 years [7]. An excellent summary of Bayesian techniques for state space models is given by [36].

In order to proceed, two models are required: first, the measurement model relates the noisy measurements to the state; and second, the system or dynamic model describes the evolution of the state with time. The measurement model used for BTO and BFO metadata is defined in a probabilistic form in Chap. 5. The dynamic model used to define the behaviour of the aircraft is defined in Chaps. 6 and 7.

If the measurement model and the system model are both linear and Gaussian, the optimal estimate can be calculated in closed form using the Kalman filter [25]. If either the system or measurement model is nonlinear or non-Gaussian, the posterior pdf will be non-Gaussian and standard analysis with a Kalman filter will be suboptimal. This results in the need for approximate computational strategies and the approach adopted in this study is introduced in this chapter. The application of the measurement and dynamics models to this approach is described in Chap. 8. The computational approach proceeds in essentially two stages: prediction and update. The prediction stage uses the aircraft dynamic model to step from the state pdf at one time to the pdf at the next time. The state is subject to unknown disturbances, modeled as random noise, and also unknown control inputs, such as turn commands, and so prediction generally translates, deforms, and broadens the state pdf. The update operation uses the latest measurement to modify (typically to tighten) the prediction pdf. This is achieved using Bayes theorem, (3.3), which is the mechanism for updating knowledge about the state in the light of extra information from new data.

3.1   The Problem and its Conceptual Solution

To define the problem of nonlinear filtering, let us introduce the state vector $\mathbf{x}(t) \in \mathbb{R}^n$, where n is the dimension of the state vector. Here t is continuous-valued time. The state evolution is best described using a continuous-time stochastic differential equation, sometimes specifically referred to as an Itô differential equation [23]. However, it is often more convenient to sample this at discrete time instants, in which case $\mathbf{x}_k \equiv \mathbf{x}(t_k)$ represents the state at the kth discrete sample time. The elapsed time between samples $\Delta_k = t_k - t_{k-1}$ is not necessarily constant. The state is assumed to evolve according to a continuous-time stochastic model:

$$\mathrm{d}\mathbf{x}(t) = \mathbf{f}\left(\mathbf{x}(t), \mathrm{d}\mathbf{v}(t), t, \mathrm{d}t\right), \tag{3.4}$$

where $\mathbf{f}(\cdot)$ is a known, possibly nonlinear deterministic function of the state and $\mathbf{v}(t)$ is referred to as a process noise sequence, which caters for random disturbances in the aircraft motion.

A sensor collects measurements, which are a possibly nonlinear function of the state. Measurements occur at times t_k, for $k = \{1, 2, \ldots K\}$. The kth measurement is denoted $\mathbf{z}_k \in \mathbb{R}^m$ where m is the dimension of the measurement vector. The measurements are related to the state via the measurement equation:

$$\mathbf{z}_k = \mathbf{h}_k\left(\mathbf{x}_k, \mathbf{w}_k\right), \tag{3.5}$$

where $\mathbf{h}_k(\cdot)$ is a known, possibly nonlinear function and $\mathbf{w}_k$ is a measurement noise sequence. The noise sequences $\mathbf{v}(t)$ and $\mathbf{w}_k$ will be assumed to be white, with known probability density functions and mutually independent. The initial state is assumed to have a known pdf $p(\mathbf{x}_0)$ and also to be independent of noise sequences.

We seek estimates of $\mathbf{x}_k$ based on the sequence of all available measurements up to time t_k, defining the measurement history $\mathbf{Z}_k \triangleq \{\mathbf{z}_1, \ldots \mathbf{z}_k\}$. From a Bayesian perspective, the problem is to recursively construct the posterior pdf $p(\mathbf{x}_k | \mathbf{Z}_k)$. In principle, the pdf $p(\mathbf{x}_k | \mathbf{Z}_k)$ may be obtained recursively in two stages: prediction and update. The prediction stage steps from the pdf of $\mathbf{x}$ at time t_{k-1}, $p(\mathbf{x}_{k-1} | \mathbf{Z}_{k-1})$, to the pdf at the next time, $p(\mathbf{x}_k | \mathbf{Z}_{k-1})$, not incorporating any new measurements. The update stage takes the predicted pdf $p(\mathbf{x}_k | \mathbf{Z}_{k-1})$ and incorporates the new measurement $\mathbf{z}_k$ occurring at time t_k to obtain the updated pdf $p(\mathbf{x}_k | \mathbf{Z}_k)$. If there is a requirement to evaluate the pdf at time t for which there is no measurement then this pdf is the predicted pdf and no update step needs to be performed.

3.1.1   Prediction

The prediction stage involves using the system model (3.4) to obtain the prediction density of the state at time step k via the Chapman–Kolmogorov equation:

$$p\left(\mathbf{x}_k | \mathbf{Z}_{k-1}\right) = \int p\left(\mathbf{x}_k | \mathbf{x}_{k-1}, \mathbf{Z}_{k-1}\right) p\left(\mathbf{x}_{k-1} | \mathbf{Z}_{k-1}\right) d\mathbf{x}_{k-1},$$

$$= \int p\left(\mathbf{x}_k | \mathbf{x}_{k-1}\right) p\left(\mathbf{x}_{k-1} | \mathbf{Z}_{k-1}\right) d\mathbf{x}_{k-1}. \tag{3.6}$$

The first line of (3.6) is a statement of the law of total probability. The simplification $p\left(\mathbf{x}_k | \mathbf{x}_{k-1}, \mathbf{Z}_{k-1}\right) = p\left(\mathbf{x}_k | \mathbf{x}_{k-1}\right)$ used to progress from the first line of (3.6) to the second applies because (3.4) describes a Markov process of order one. The probabilistic model of the state evolution, $p\left(\mathbf{x}_k | \mathbf{x}_{k-1}\right)$, is defined by the system equation (3.4) and the known statistics of $\mathbf{v}(t)$.

3.1.2   Update

At time t_k a measurement $\mathbf{z}_k$ becomes available and the update stage is carried out. This involves an update of the prediction (or prior) pdf via Bayes' rule:

$$p\left(\mathbf{x}_k | \mathbf{Z}_k\right) = p\left(\mathbf{x}_k | \mathbf{z}_k, \mathbf{Z}_{k-1}\right)$$

$$= \frac{p\left(\mathbf{z}_k | \mathbf{x}_k, \mathbf{Z}_{k-1}\right) \ p\left(\mathbf{x}_k | \mathbf{Z}_{k-1}\right)}{p\left(\mathbf{z}_k | \mathbf{Z}_{k-1}\right)}$$

$$= \frac{p\left(\mathbf{z}_k | \mathbf{x}_k\right) \ p\left(\mathbf{x}_k | \mathbf{Z}_{k-1}\right)}{p\left(\mathbf{z}_k | \mathbf{Z}_{k-1}\right)}, \tag{3.7}$$

where conditional independence has been used to write the likelihood function $p\left(\mathbf{z}_k | \mathbf{x}_k, \mathbf{Z}_{k-1}\right) = p\left(\mathbf{z}_k | \mathbf{x}_k\right)$, which is defined by the measurement model (3.5) and the known statistics of $\mathbf{w}_k$. The normalizing constant on the denominator can be expanded as

$$p\left(\mathbf{z}_k | \mathbf{Z}_{k-1}\right) = \int p\left(\mathbf{z}_k | \mathbf{x}_k\right) \ p\left(\mathbf{x}_k | \mathbf{Z}_{k-1}\right) d\mathbf{x}_k. \tag{3.8}$$

In the update stage (3.7), the measurement $\mathbf{z}_k$ is used to modify the prior density to obtain the required posterior density of the current state.

Note that there is no requirement for all of the measurements to have the same statistical model or even contain the same type of information. For example, there could be multiple sensors operating on different modalities. For simplicity, we have not introduced explicit notation to change the measurement pdf for each k. For the accident flight three different types of measurement have been used. As discussed in Chap. 5, the satellite communications messages consist of R-channel and C-channel messages that have differing information content. Another quite different form of measurement is the areas of the ocean floor that have been searched without locating the aircraft and the debris that has been recovered. This measurement and its potential use to refine the ongoing search are discussed in Chap. 11.

The recurrence relations (3.6) and (3.7) form the basis for the optimal Bayesian solution. The recursive propagation of the posterior density, given by (3.6) and (3.7), is only a conceptual solution in the sense that in general it cannot be determined analytically. In most practical situations the analytic solution of (3.7) and (3.8) is intractable and numerical approximations have to be used. This has been a topic of significant research effort over the past 20 years [1, 20, 33]; a general overview of the method is presented next.

3.2 The Particle Filter

In the linear Gaussian case, the pdfs for $p(\mathbf{v}(t))$, $p(\mathbf{w}_k)$, $p(\mathbf{x}_0)$ are all Gaussian and the functions $\mathbf{f}(\cdot)$ and $\mathbf{h}(\cdot)$ are linear. It can then be easily shown that the posterior $p(\mathbf{x}_k|\mathbf{Z}_k)$ is also Gaussian and all of these pdfs can be summarised completely by their means and covariances. The Kalman filter is an algorithm that defines recursions for the mean and covariance of $p(\mathbf{x}_k|\mathbf{Z}_k)$ in terms of the means and covariances of the prior and noise processes. However, in general, the posterior does not take the same functional form as the prior and indeed it is not possible to even write a closed form expression for $p(\mathbf{x}_k|\mathbf{Z}_k)$. In this case an approximate solution is required. The solution used for the MH370 search definition is referred to as the particle filter and is a numerical approximation based on random sampling.

The fundamental concept in the particle filter is to approximate the pdf $p(\mathbf{x}_k|\mathbf{Z}_k)$ as a weighted combination of sample points

$$p(\mathbf{x}_k|\mathbf{Z}_k) \approx \sum_{p=1}^{P} w_k^p \delta\left(\mathbf{x}_k - \mathbf{x}_k^p\right),\tag{3.9}$$

where the w_k^p are referred to as weights and sum to unity, and the $\mathbf{x}_k^p$ are referred to as particles. The convergence properties of this approximation in the limit as the number of particles P increases have been well studied, for example [14, 21]. Given this approximate pdf, it is simple to evaluate the expectation of any nonlinear function of the state, such as

$$\mathbb{E}\left[\mathbf{g}(\mathbf{x}_k)|\mathbf{Z}_k\right] \equiv \int \mathbf{g}(\mathbf{x}_k)\, p(\mathbf{x}_k|\mathbf{Z}_k)\, d\mathbf{x}_k \approx \sum_{p=1}^{P} w_k^p \mathbf{g}\left(\mathbf{x}_k^p\right).\tag{3.10}$$

The approximation of an integral using sample points as above is referred to as Monte Carlo integration and can be applied to both the Chapman–Kolmogorov prediction (3.6) and the Bayesian update (3.7).

The particle filter is an algorithm that provides a mechanism to recursively create a set of weighted particles approximating $p(\mathbf{x}_k|\mathbf{Z}_k)$ starting from a previous set of weighted particles approximating $p(\mathbf{x}_{k-1}|\mathbf{Z}_{k-1})$. It does this in two stages: first it moves the particle sample points $\mathbf{x}_{k-1}^p \rightarrow \mathbf{x}_k^p$ to new locations using a pdf referred to as a *proposal* distribution, which is a tractable approximation of the pdf of interest. Second, it determines new particle weights to correct for the difference between the proposal and the true pdf. This process is known as importance sampling [1, 33].

The proposal distribution is a critical component of the particle filter. It is a function chosen by the designer subject to relatively loose constraints. Importantly, the proposal distribution must cover all of the state space where the true distribution is non-zero and its tails should be heavier than the tails of the true distribution. If the proposal is chosen poorly then many of the particles $\mathbf{x}_k^p$ will be assigned very low weights and the filter efficiency will be low: a large number of particles will be required for satisfactory performance. A common version of the particle filter is the Sample-Importance-Resample (SIR) particle filter that uses the system dynamics as a proposal distribution. The SIR is popular because it is often relatively straightforward to sample from the dynamics and because the weight update equation is very simple when the dynamics is used as the proposal. The filter used in this book is a form of SIR particle filter.

For the SIR particle filter, for each particle $\mathbf{x}_{k-1}^p$ a new $\mathbf{x}_k^p$ is drawn from the transition density $p(\mathbf{x}_k | \mathbf{x}_{k-1}^p)$, and weights are updated by scaling the previous weights by the current measurement likelihood and re-normalising,

$$w_k^p = (W_k)^{-1} \, p\left(\mathbf{z}_k | \mathbf{x}_k^p\right) w_{k-1}^p, \tag{3.11}$$

where the normalising term is $W_k = \sum_{p=1}^P p\left(\mathbf{z}_k | \mathbf{x}_k^p\right) w_{k-1}^p$.

A key difficulty in particle filters is the issue of degeneracy, i.e., over time, many weights tend toward zero, and the corresponding particles are of little use. Resampling is used to combat this difficulty. The simplest approach is draw P new particles from the approximate distribution (3.9), such that particles with very large weights are likely to be replicated many times over, and those with very small weights are unlikely to be sampled. A variety of methods are possible, and can be found in [1, 33]. The sampling method used in this study is detailed in Chap. 8.

3.3   Rao–Blackwellised Particle Filter

One of the challenges in implementing a particle filter is that the number of particles required to make a good approximation to the desired posterior pdf can grow exponentially with the dimension of the state space. In some circumstances, it is possible to mitigate this by incorporating an analytic representation of the distribution of part of the state given a sample of the remainder of the state. For example, suppose that the measurement function can be decomposed into two parts

$$\mathbf{z}_k = \mathbf{h}_k^1\left(\mathbf{x}_k^1\right) + \mathbf{h}_k^2\left(\mathbf{x}_k^2\right) + \mathbf{w}_k, \tag{3.12}$$

where $\mathbf{x}_k^1$ and $\mathbf{x}_k^2$ are disjoint sub-vectors of the state $\mathbf{x}_k$. In this case we can write

$$p\left(\mathbf{x}_k^1, \mathbf{x}_k^2 | \mathbf{z}_t, \mathbf{Z}_{t-1}\right) = p\left(\mathbf{x}_k^1 | \mathbf{z}_t, \mathbf{Z}_{t-1}\right) p\left(\mathbf{x}_k^2 | \mathbf{x}_k^1, \mathbf{z}_t, \mathbf{Z}_{t-1}\right). \qquad (3.13)$$

The two densities above can be estimated using different filters. When the function $\mathbf{h}_k^2\left(\mathbf{x}_k^2\right)$ is linear and the noise is Gaussian, the second density $p\left(\mathbf{x}_k^2 | \mathbf{x}_k^1, \mathbf{z}_t, \mathbf{Z}_{t-1}\right)$ can be estimated using a Kalman filter, even if the first function $\mathbf{h}_k^1\left(\mathbf{x}_k^1\right)$ is nonlinear. The state vector that needs to be sampled is then $\mathbf{x}_k^1$ not $[\mathbf{x}_k^1, \mathbf{x}_k^2]$ and the sampling process can use fewer samples for a given degree of accuracy.

When a particle filter is used for the nonlinear part of the measurement problem, the conditioning of the second state density $p\left(\mathbf{x}_k^2 | \mathbf{x}_k^1, \mathbf{z}_t, \mathbf{Z}_{t-1}\right)$ leads to a separate Kalman filter for each particle. Each Kalman filter uses the sampled value of the sub-state $\mathbf{x}_k^1$ as though it were the truth. This arrangement is referred to as a Rao–Blackwellised particle filter [15, 29, 38].

Open Access This chapter is distributed under the terms of the Creative Commons Attribution-NonCommercial 4.0 International License (http://creativecommons.org/licenses/by-nc/4.0/), which permits any noncommercial use, duplication, adaptation, distribution and reproduction in any medium or format, as long as you give appropriate credit to the original author(s) and the source, a link is provided to the Creative Commons license and any changes made are indicated.

The images or other third party material in this chapter are included in the work's Creative Commons license, unless indicated otherwise in the credit line; if such material is not included in the work's Creative Commons license and the respective action is not permitted by statutory regulation, users will need to obtain permission from the license holder to duplicate, adapt or reproduce the material.

Chapter 4
Aircraft Prior Based on Primary Radar Data

The Bayesian approach described in the previous chapter is a recursive method that calculates the posterior state distribution at each measurement time from a distribution at the previous measurement time. It fundamentally requires knowledge of three probability density functions: the prior distribution of the state at initialisation, $p(\mathbf{x}(0))$; the state evolution $p(\mathbf{x}_k|\mathbf{x}_{k-1})$; and the measurement likelihood $p(\mathbf{z}_k|\mathbf{x}_k)$. Chapter 5 addresses the measurement probability density and Chaps. 6 and 7 discuss the state transition model. This chapter discusses the prior state distribution and the method used to define it. Intuitively, one might expect this prior to have a significant influence on the probability distribution at later times: a larger uncertainty in the prior might be expected to lead to a greater spread of uncertainty in the final pdf compared to a prior pdf with smaller uncertainty.

In the MH370 search, there are two data sources that are available to construct the prior. The aircraft reports its own location and other information to the ground via a satellite link using the Aircraft Communications Addressing and Reporting System (ACARS). Data from ACARS is available for the accident flight only up to the point where communications were lost: the final ACARS report was at 17:07:29. In Chap. 9, other flights with known aircraft locations are used to validate the models used for the accident flight. For these, ACARS data is available and this data is used to construct the prior.

The second source of prior information is radar. For the validation flights this radar data is not available and nor is it required given the presence of ACARS logs. For the accident flight, primary radar data provided by Malaysia is available from after the loss of communications up until 18:22:12. The radar data contains regular estimates of latitude, longitude and altitude at 10 s intervals from 16:42:27 to 18:01:49. A single additional latitude and longitude position was reported at 18:22:12. Figure 4.1 shows the radar data overlaid on a map. Under radar coverage, the aircraft turned sharply at approximately 17:24, crossed over Malaysia, and then turned to the North-West at 17:53.

The Bayes filter requires a prior over the full state space, not merely latitude and longitude. The development to follow will lead to a state vector containing several other parameters. Where possible it is preferable to specify a prior on these parameters

© Commonwealth of Australia 2016
S. Davey et al., *Bayesian Methods in the Search for MH370*, SpringerBriefs
in Electrical and Computer Engineering, DOI 10.1007/978-981-10-0379-0_4

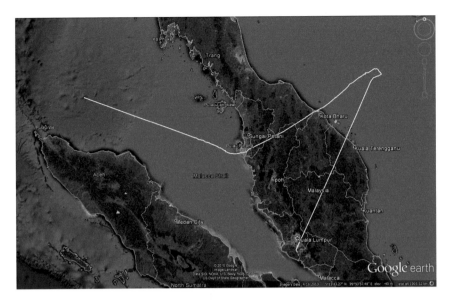

Fig. 4.1 Primary and secondary radar data available for MH370

using radar data or ACARS rather than subjective belief. Where this is not possible, the philosophy has been to use priors that are diffuse to avoid prejudicing the filter output. It is possible to derive the angle and speed of the ground velocity from the radar reports by assuming a simplified almost constant velocity model and applying a Kalman filter. This assumption is acceptable for the primary radar data because the reports are closely spaced in time. Figure 4.2 shows the derived speed and direction obtained from this filter.[1] The speed estimates vary dramatically during the first turn, which is not an accurate representation of the aircraft speed at this time. It is likely due to the mismatch between the assumed linear Kalman filter model and the high acceleration manoeuvre performed by the aircraft. Since these artefacts are localised to the time of the turn the influence on the state at the end of the sequence is negligible.

The final reported position from radar was at very long range from the sensor and there was a long time delay between it and the penultimate radar report. This report is at long range and it is likely to have rather poor accuracy because the angular errors translate to large location errors at long range. The radar report at 18:22 is closer to the penultimate report at 18:02 than the filter speed predicts. Also, it was observed that the range ring derived from the timing measurements at 18:25 and 18:28 are closer to the 18:02 report than predictions based on either the 18:02 filtered speed or the 18:22 filtered speed. Figure 4.3 shows the relative positions of the 18:25 arc and the filter predictions based on data up to 18:02. Collectively these data points suggest that the aircraft may have slowed down at some point between 18:02 and 18:22. In

---

[1]The measurement error was assumed to have a standard deviation of 0.5 nm and the process noise variance was 3.5×10^{-4} nm^2 s^{-3}. The process noise was adjusted to minimise the mean squared prediction error.

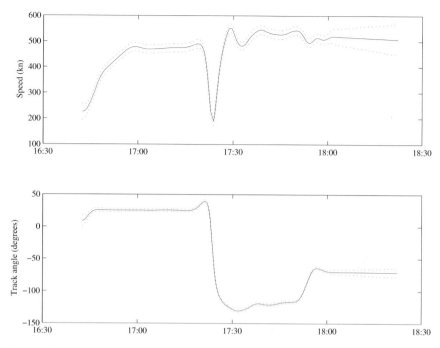

Fig. 4.2 Smoothed estimates of speed and direction derived from radar data. *Dotted lines* show covariance of estimates, illustrated as mean plus and minus three-sigma value

addition, the ground speed observed by the radar prior to 18:02 is relatively high and implies that the aircraft would have been at low altitude. This is likely to result in poor fuel efficiency, and in order to maintain flight for the duration indicated by the satellite data, the aircraft would have had to slow and increase altitude at some stage to conserve fuel. This is also consistent with a potential speed change between 18:02 and 18:22.

The 18:22 radar observation was not used quantitatively because the latitude and longitude derived from it are likely to be less accurate at long range and the aircraft may have manoeuvred prior to 18:22. The radar observation was deemed to indicate that the aircraft did not turn between 18:02 and 18:22, but the numerical values were not used. Instead, a prior was defined at 18:01 at the penultimate radar point using the output of the Kalman filter described above. The position standard deviations were set to 0.5 nm and the direction standard deviation to 1°. Figure 4.3 shows predictions of the mean of this prior from 18:02 to 18:25, shown in yellow, and one-sigma lines at ±1°. The 18:22 radar point, at the end of the radar track, is clearly within the azimuth fan. As described above, the filtered speed at the output of the Kalman filter is not consistent with the 18:25 measurement and predictions based purely on this will have a likelihood very close to zero. In addition, the model discussed in Chap. 6 specifies air speed in terms of Mach number. The manoeuvre model described in Chap. 7 allows for speed changes and these will be randomly sampled by the proposal distribution.

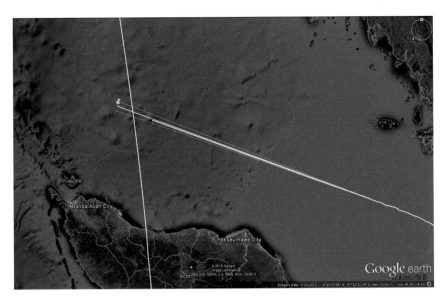

Fig. 4.3 18:02 prediction to 18:25, shown in *yellow*. The Malaysian military radar track is shown in *white*, on the *right*. The near-*vertical white line* on the *left* corresponds to the 18:25 BTO arc

Rather than trying to specify when the speed change occurred, the filter was expected to learn this information. This provides a richer description of the trajectory since the timing of a speed change and the new speed selected are coupled together to arrive at an appropriate position at 18:25. As will be seen, the filter had no difficulty finding paths that agreed with the measurement data. An initial Mach number was selected from a uniform prior between 0.73 and 0.84; this was chosen on the basis of expert advice to ensure that the required flight endurance is achievable.

It is possible to define a much earlier prior using only the ACARS data from early in the flight. In this case, the turns illustrated in the radar data become part of the unknown aircraft flight path to be estimated. Chapter 10 illustrates that this approach leads to a larger search zone, including the search zone resulting from the use of the radar data, and extending further North. This broadening occurs because the flight path from the final ACARS report to 18:25 is much less constrained.

Open Access   This chapter is distributed under the terms of the Creative Commons Attribution-NonCommercial 4.0 International License (http://creativecommons.org/licenses/by-nc/4.0/), which permits any noncommercial use, duplication, adaptation, distribution and reproduction in any medium or format, as long as you give appropriate credit to the original author(s) and the source, a link is provided to the Creative Commons license and any changes made are indicated.

The images or other third party material in this chapter are included in the work's Creative Commons license, unless indicated otherwise in the credit line; if such material is not included in the work's Creative Commons license and the respective action is not permitted by statutory regulation, users will need to obtain permission from the license holder to duplicate, adapt or reproduce the material.

Chapter 5
Measurement Model, Satellite Communications

The Bayesian filter discussed in Chap. 3 relies on knowledge of three probability density functions: the state prior distribution, the state stochastic model, and the measurement conditional probability density. The prior used for the analysis in this book was discussed in Chap. 4. This chapter addresses the measurement probability density and Chaps. 6 and 7 discuss the state dynamics model.

The most general measurement model was defined in (3.5) and simply states that the measurement is some function of the aircraft state and measurement noise. In many systems it is not too restrictive to assume that the noise is additive, in which case (3.5) becomes

$$\mathbf{z}_k = \mathbf{h}_k(\mathbf{x}_k) + \mathbf{w}_k. \tag{5.1}$$

Prescribing the nonlinear function $\mathbf{h}_k(\mathbf{x}_k)$ and the statistical distribution of the measurement noise provides a complete description of the measurement probability density. The measurements available for the accident flight are timing and frequency logs of communication messages between the aircraft and a ground station. Details of the communication system software and hardware combine with the physics of the communication geometry to determine the nonlinear measurement function. The statistics of the noise were determined empirically by analysing a population of real measurements for known aircraft states.

This chapter gives a brief overview of the satellite communication system and then describes the nonlinear measurement functions and empirical noise models for the timing and frequency measurements. The majority of the communications messages available were automated signalling messages, but there were also two telephone calls made to the plane that remained unanswered. The first of these is particularly important because of when it occurred. The chapter concludes with a description of the measurement model for telephony. Further details may be found in [2].

© Commonwealth of Australia 2016
S. Davey et al., *Bayesian Methods in the Search for MH370*, SpringerBriefs
in Electrical and Computer Engineering, DOI 10.1007/978-981-10-0379-0_5

5.1   Satellite Communications System

The accident aircraft was fitted with a satellite communications terminal that used the Inmarsat Classic Aero system [2]. This system uses a satellite to relay messages between the aircraft and a ground station. In the accident flight the messages were passed between the aircraft and a ground receiving unit located in Perth, Australia, via the Inmarsat-3F1 satellite. Figure 5.1 illustrates the satellite communication system in use during the accident flight. The aircraft is referred to as the Aircraft Earth Station (AES) and the ground receiving unit is referred to as the Ground Earth Station (GES). Inmarsat-3F1 is a satellite in geosynchronous orbit with longitude 64.5° East and was used exclusively for the duration of the accident flight.

An AES is equipped with a satellite data unit that comprises a satellite modem and auxiliary hardware and software. Transmission of data over the satellite is via bursts which are scheduled to arrive at the GES at a specified time and frequency. Communications from multiple users are coordinated by the allocation of different time and frequency slots to each user. Return channel (AES to GES) time slot boundaries are referenced to the forward channel (GES to AES) [2]. The duration of each time slot is sufficient to account for all possible positions of the aircraft with respect to the satellite. The width of each frequency slot is determined by the data rate and a guard width that accounts for possible variations in the satellite oscillator frequency and other possible frequency offsets. Frequency compensations applied onboard the aircraft (aircraft induced Doppler pre-compensation) and at the ground station serve to reduce the possible difference between the expected and actual frequency of the

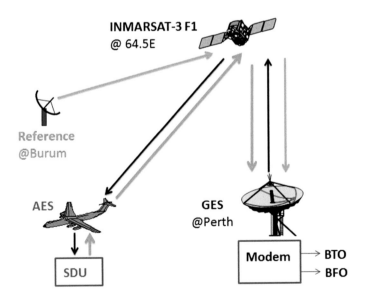

Fig. 5.1 System model of the satellite communication system

messages received from the aircraft. The on-ground compensation makes use of a second ground station located in Burum, Netherlands that transmits a reference signal to the Inmarsat-3F1 satellite which is relayed to the Perth GES. Its purpose is to enable the receive modem in the GES to compensate for the Doppler frequency shift from the satellite to the Perth GES. This compensation process is referred to as Enhanced Automatic Frequency Correction.

After the Enhanced Automatic Frequency Correction process, the expected time of arrival of each communications burst is compared with the actual time of arrival and the difference between the two is referred to as the Burst Timing Offset (BTO). The BTO is minimised when the elevation angle to the satellite is 90° and increases as the aircraft moves away from the sub-satellite position. Hence, the BTO is a measure of how far the aircraft is from the sub-satellite position. Similarly, the difference between the expected frequency of each communications burst and the actual received frequency is referred to as the Burst Frequency Offset (BFO). The BFO is a function of the relative speed between the aircraft and the satellite. Given that the satellite position and speed are known, the BFO provides information about the aircraft velocity vector. The BTO and BFO are logged by the ground station for every communications burst. This logging was a relatively recent addition to the ground station following the Air France 447 accident [2, 45] and was intended to assist in locating an aircraft. Statistical models for these two measurement functions are now developed.

5.2 Burst Timing Offset

The Inmarsat Classic Aero system allocates a time slot for communications based on a nominal propagation delay that assumes a nominal satellite position and a nominal aircraft position. The nominal aircraft position is at zero altitude directly below the satellite's nominal orbital position of 64.5°E longitude, zero latitude and an altitude of 35788.122 km. The round trip delay is proportional to the distance from the ground station to the actual aircraft location via the actual satellite location. The actual satellite position is different from the nominal satellite position because Inmarsat-3F1 is not exactly motionless, but rather moves in a known way in a region about its nominal location. The Burst Timing Offset is the additional delay after the start of the allocated time slot at which the message is received [2]. The BTO is thus the difference between the round trip message delay and the nominal delay used for scheduling. In addition to the propagation delay the message delay includes the latency of the satellite data processing unit. Denoting a BTO measurement at time t_k as z_k^{BTO}, the BTO measurement function is

$$z_k^{BTO} = h_k^{BTO} (\mathbf{x}_k, \mathbf{s}_k) + w_k^{BTO}, \tag{5.2}$$

$$h_k^{BTO} (\mathbf{x}_k, \mathbf{s}_k) = T (\mathbf{x}_k, \mathbf{s}_k) - T^{nom} + T^{channel} - T_k^{anomaly}, \tag{5.3}$$

where

- $T\,(\mathbf{x}_k, \mathbf{s}_k)$ is the round trip propagation delay from the ground station to the aircraft via the satellite;
- $\mathbf{s}_k$ is the state of the satellite at time t_k, that is its position and velocity in three dimensions, along with the satellite oscillator's state;
- T^{nom} is the nominal round trip delay;
- T^{channel} is a channel dependent bias term due to processing in the satellite data unit;
- T_k^{anomaly} is an anomaly correction term discussed below;
- w_k^{BTO} is a zero mean scalar noise process with statistics to be determined from measurement data logs.

The function $h_k^{\mathrm{BTO}}\,(\mathbf{x}_k, \mathbf{s}_k)$ is assumed to be deterministic, so the measurement variance is the variance of the noise term w_k^{BTO}. The round trip delay can be expressed in terms of the distances from the satellite to the ground station and the aircraft as

$$T\,(\mathbf{x}_k, \mathbf{s}_k) = \frac{2}{c}\Big(\big|\mathbf{H}^s \mathbf{s}_k - \mathbf{g}\big| + \big|\mathbf{H}^s \mathbf{s}_k - \mathbf{H}^x \mathbf{x}_k\big|\Big), \qquad (5.4)$$

where c is the speed of light; $\mathbf{g}$ is the location of the ground station; $\mathbf{H}^s$ selects the position elements of the satellite state; $\mathbf{H}^x$ selects the position elements of the aircraft state; and the distance $|\cdot|$ is the three dimensional Cartesian distance, i.e. the l^2 norm.

Combining the nominal locations of the satellite and aircraft with the known location of the Perth ground station gives a value of $T^{\mathrm{nom}} = 499{,}962\,\mu\mathrm{s}$.

There are a number of different channel types used that carry different traffic types and have different baud rates. Communications from the aircraft to the ground are typically over the R- and T-data channels with C-channel used for voice telephone calls. Communications from the ground to the aircraft are over the P-channel. The channel dependent calibration term T^{channel} is assumed to be constant over a single flight but can vary between flights. A fixed value for each flight assessed was empirically derived by comparing the communications logs with known aircraft positions: the calculated value of T^{channel} was the mean difference between the measured BTO and the expected BTO calculated using the known aircraft location. For the accident flight this calibration is only available for the time when the plane was at the tarmac and for the first half hour of flight. As such, values from the previous flight were also used in the calculation of T^{channel}. The majority of the messages available from the accident flight are R1200 messages for which $T^{\mathrm{channel}} = -4{,}283\,\mu\mathrm{s}$.

The anomaly correction term T_k^{anomaly} was empirically derived through analysis of a collection of communications logs. For some communication messages, typically during initial log-on, there was a very large difference between the measured BTO and the nominal delay. Analysis showed that rather than simple outliers, these anomalous BTO measurements could be *corrected* by a factor of $N \times 7{,}820\,\mu\mathrm{s}$ where N is a positive integer. The origin of these anomalous BTO measurements has not been fully determined, but the empirical correction time is quite close to the transmission interrupt clock period of 7,812.5 µs and the BTO collection process contains quantisation.

The channel dependent calibration and anomaly correction terms result in a residual measurement error that is approximately Gaussian. For the R1200 messages, the empirically derived standard deviation of the measurement noise w_k^{BTO} is 29 µs, and for R600 messages, 62 µs. For anomalous R1200 messages a standard deviation of 43 µs was used. Figure 5.2 shows a histogram of the residual BTO measurement errors for R1200 messages referenced to the 7 March 2014 $T^{channel}$ value, and the moment-matched Gaussian approximation. The data used to construct the histogram and the empirical parameters were obtained from logs of the 20 flights of 9M-MRO prior to the accident flight.

The histogram in Fig. 5.2 has an underlying mean of 10 µs. This is due to the channel dependent calibration term $T^{channel}$ not being stationary. Over the span of a

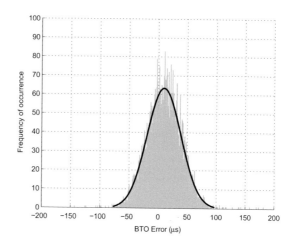

Fig. 5.2 Histogram of BTO residual measurement errors

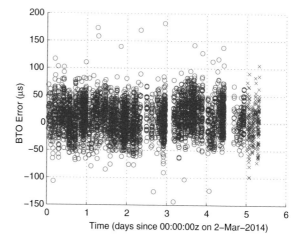

Fig. 5.3 Scatter plot of BTO errors for 20 flights prior to MH370

day it appears constant but in the context of the 20 flights represented in Fig. 5.2 there is a slow variation. As discussed above, different values were fitted for each flight. Figure 5.3 shows a scatter plot of the BTO errors against time. The channel dependent calibration term T^{channel} was matched to the final flight before the accident flight, MH371 (and the beginning of the accident flight) and the BTO errors from MH371 on 7 March 2014 are marked as red crosses. The variation in bias is sufficiently slow that assuming it is the same for the accident flight as the previous flight is satisfactory.

5.3 Burst Frequency Offset

The Burst Frequency Offset is a function of the Doppler shifts imparted on the communication signal due to the motion of the satellite and the aircraft. The relationship is more complicated than a direct Doppler calculation because the aircraft software contains Doppler compensation that offsets the Doppler shift due to the aircraft motion. Although the aircraft attempts to compensate for its own motion, it does this under the assumption that the communications satellite is in motionless geostationary orbit and it does not include the vertical component of the aircraft velocity (which is non-zero when it is ascending or descending) [2]. Since Inmarsat-3F1 is not exactly geostationary, the compensation is unable to completely remove Doppler effects. Empirical analysis of the BFO was conducted for the 20 flights of 9M-MRO prior to the accident flight. This analysis used the same Doppler correction software as the 9M-MRO satellite data unit to determine the expected BFO given a known reported aircraft position and velocity and compared this with the observed measurements.

Similar to the BTO, the BFO measurement function consists of a nonlinear function of the aircraft and satellite states and several bias terms [2]

$$z_k^{\text{BFO}} = h_k^{\text{BFO}}(\mathbf{x}_k, \mathbf{s}_k) + w_k^{\text{BFO}} \tag{5.5}$$
$$h_k^{\text{BFO}}(\mathbf{x}_k, \mathbf{s}_k) = \Delta F_k^{\text{up}}(\mathbf{x}_k, \mathbf{s}_k) + \Delta F_k^{\text{down}}(\mathbf{s}_k) + \delta f_k^{\text{comp}}(\mathbf{x}_k) + \delta f_k^{\text{sat}}(\mathbf{s}_k)$$
$$+ \delta f_k^{\text{AFC}}(\mathbf{s}_k) + \delta f_k^{\text{bias}}(\mathbf{x}_k, \mathbf{s}_k), \tag{5.6}$$

where

- $\Delta F_k^{\text{up}}(\mathbf{x}_k, \mathbf{s}_k)$ is the uplink (aircraft to satellite) Doppler shift;
- $\Delta F_k^{\text{down}}(\mathbf{s}_k)$ is the downlink (satellite to ground station) Doppler shift;
- $\delta f_k^{\text{comp}}(\mathbf{x}_k)$ is the frequency compensation applied by the aircraft;
- $\delta f_k^{\text{sat}}(\mathbf{s}_k)$ is the variation in satellite translation frequency: the satellite uses a local oscillator to translate the carrier frequency of the message;
- $\delta f_k^{\text{AFC}}(\mathbf{s}_k)$ is the frequency compensation applied by the ground station receive chain;
- $\delta f_k^{\text{bias}}(\mathbf{x}_k, \mathbf{s}_k)$ is a slowly varying bias due to errors in the aircraft and satellite oscillators and processing in the satellite data unit;
- w_k^{BFO} is a zero mean scalar noise process with statistics to be determined from measurement data logs.

This function was described in detail in [2], we review it briefly and elaborate where the analysis herein makes different modeling assumptions to those in [2]. Again, the function $h_k^{\text{BFO}}(\mathbf{x}_k, \mathbf{s}_k)$ is assumed to be deterministic, so the measurement variance is the variance of the noise term w_k^{BFO}. This is a less reliable assumption than for BTO because the bias term $\delta f_k^{\text{bias}}(\mathbf{x}_k, \mathbf{s}_k)$ changes. To compensate for this, the measurement variance was inflated from the empirically derived w_k^{BFO} variance.

The uplink Doppler shift can be expressed as a product of the uplink frequency with the inner product of the relative velocity (normalised by the speed of light, c) and a unit vector along the relative displacement between the aircraft and the satellite

$$\Delta F_k^{\text{up}}(\mathbf{x}_k, \mathbf{s}_k) = \frac{F^{\text{up}}}{c} \frac{(\mathbf{V}^s \mathbf{s}_k - \mathbf{V}^x \mathbf{x}_k)^\top (\mathbf{H}^s \mathbf{s}_k - \mathbf{H}^x \mathbf{x}_k)}{|\mathbf{H}^s \mathbf{s}_k - \mathbf{H}^x \mathbf{x}_k|}, \tag{5.7}$$

where F^{up} is the uplink carrier frequency; $\mathbf{V}^s$ selects the velocity elements of the satellite state; and $\mathbf{V}^x$ selects the velocity elements of the aircraft state. Similarly the downlink Doppler shift and the frequency compensation are given by

$$\Delta F_k^{\text{down}}(\mathbf{s}_k) = \frac{F^{\text{down}}}{c} \frac{(\mathbf{V}^s \mathbf{s}_k)^\top (\mathbf{H}^s \mathbf{s}_k - \mathbf{g})}{|\mathbf{H}^s \mathbf{s}_k - \mathbf{g}|}, \tag{5.8}$$

$$\delta f_k^{\text{comp}}(\mathbf{x}_k) = \frac{F^{\text{up}}}{c} \frac{(\bar{\mathbf{V}}^x \mathbf{x}_k)^\top (\bar{\mathbf{H}}^x \mathbf{x}_k - \bar{\mathbf{s}})}{|\bar{\mathbf{H}}^x \mathbf{x}_k - \bar{\mathbf{s}}|}, \tag{5.9}$$

where $\bar{\mathbf{s}}$ is the nominal satellite position assumed by the aircraft, and F^{down} is downlink carrier frequency. The aircraft frequency compensation term $\delta f_k^{\text{comp}}(\mathbf{x}_k)$ is determined using the aircraft's own knowledge of its position and velocity but with an assumed altitude of zero and an assumed vertical speed of zero. The modified position matrix $\bar{\mathbf{H}}^x$ selects only the horizontal location variables and sets the altitude to zero, and similarly for $\bar{\mathbf{V}}^x$. The compensation also assumes a motionless satellite at its nominal satellite location of 64.5°E. The satellite altitude used in the correction is 422 km higher than the nominal 35788.12 km value.

The satellite translates the frequency of messages using a local oscillator that is maintained in a temperature-controlled enclosure to improve its stability. During eclipse periods, when the satellite passes through the Earth's shadow, the satellite temperature drops, resulting in a small variation in translation frequency [2]. An eclipse period occurred during the accident flight and some of the validation flights were also affected by eclipses. The oscillator temperature also varies with time of day as the satellite orientation to the sun changes and as the temperature control system applies its controls. All of these thermal effects are included in the term $\delta f_k^{\text{sat}}(\mathbf{s}_k)$. The specific details of the functions that define $\delta f_k^{\text{sat}}(\mathbf{s}_k)$ and $\delta f_k^{\text{AFC}}(\mathbf{s}_k)$ are proprietary of Inmarsat.

The bias term $\delta f_k^{\text{bias}}(\mathbf{x}_k, \mathbf{s}_k)$ is time varying. In the BTO case the variations in bias were slow enough to be ignored within a single flight and we were able to assume that T^{channel} was constant for each flight. This is not the case for the BFO bias term.

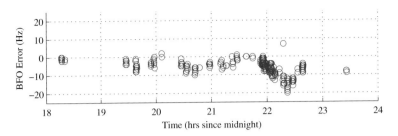

Fig. 5.4 Results for the 2-Mar-2014 flight from Mumbai to Kuala Lumpur

The mean bias is different between flights and even within a single flight there is evidence of structured variation. Figure 5.4 shows an example of the BFO measurement errors for a flight between Mumbai (BOM) and Kuala Lumpur (KUL) on 2 March 2014. The figure shows the difference between the measured BFO values and predicted values (using the actual SDU software for determining $\delta f_k^{\text{comp}}(\mathbf{x}_k)$ in the SATCOM system model) based on the known geometry and aircraft velocity vector at the time. The bias used for the plot was obtained by analysing BFO measurements while the aircraft was on the tarmac. The residual error is clearly not zero-mean, and the mean varies with time. Substantial effort was made to characterise this structured bias. It was found to have a geographic dependency but it has not been possible to determine a quantitative function to compensate for this change in bias.

The variations in bias shown in Fig. 5.4 happen over a timescale of minutes rather than hours. In the accident flight the available BFO values are generally at least an hour apart. This is a relatively long time compared with the correlation structure of the error, so the model does not use a coloured noise model for the BFO. However, the drift of the BFO bias means that it is not sufficient to assume that $\delta f_k^{\text{bias}}(\mathbf{x}_k, \mathbf{s}_k)$ will be the same in flight as on the tarmac before takeoff. The potential variations were incorporated by modeling the BFO bias as an unknown constant with a prior mean given by the tarmac value and a standard deviation of 25 Hz. Since the BFO measurement Eq. (5.6) is linear in the bias its distribution conditioned on the other states can be estimated with a Kalman filter. This is the Rao-Blackwellised particle filter described in Sect. 3.3.

Empirical statistics of the residual measurement noise w_k^{BFO} were determined using the previous 20 flights of 9M-MRO. Data points corresponding to when the aircraft was climbing or descending were excluded. Table 5.1 shows the empirical statistics of the BFO measurements for R1200 and R600 messages. The 'in-flight only' statistics show the combined effects of noise and bias variation without the influence of 'on-tarmac' outliers (potentially due to taxiing). The 'including tarmac' statistics on the other hand are also influenced by the BFO bias value applied to keep the BFO error at the source tarmac for R1200 messages close to zero. The mean BFO error was close to zero in all cases, indicating appropriate $\delta f_k^{\text{bias}}(\mathbf{x}_k, \mathbf{s}_k)$ values were chosen for each flight. The statistics show that even when outliers are discarded a standard deviation of about 4.3 Hz is applicable. As discussed above,

Table 5.1 Statistics of BFO errors for 20 flights of aircraft 9M-MRO prior to MH370

	Mean BFO error (outliers included) (Hz)	Standard deviation of BFO (outliers included) (Hz)	Mean BFO error (outliers excluded) (Hz)	Standard deviation of BFO (outliers excluded) (Hz)
Including tarmac	0.2246	4.9592	0.2745	4.0192
In-flight only	0.1079	5.4840	0.1755	4.3177

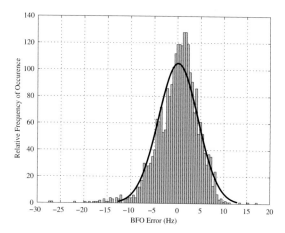

Fig. 5.5 Histogram of BFO errors for 20 flights prior to MH370 (using only in-flight BFOs)

to be conservative and allow for potential variation in the $\delta f_k^{\text{bias}}(\mathbf{x}_k, \mathbf{s}_k)$ value on the accident flight, our model assumes a noise standard deviation of 7 Hz. Section 5.5 illustrates the sensitivity of the BTO and BFO measurements to variations in the aircraft state.

Figure 5.5 shows a histogram of the 3392 in-flight BFO errors. On-tarmac BFO errors were excluded due to the pre-biasing described above. A Gaussian fit to the distribution is shown as a black line. It can be seen that the distribution shows some non-Gaussian features and the tails of the distribution for negative errors are somewhat heavier than those for positive errors.

5.4 C-Channel Telephone Calls

There were two unanswered telephone calls from the ground to MH370 after the loss of radar data. These communications use the C-channel and result in measurements of BFO but not BTO. Initially the C-channel data was not included in the flight prediction but analysis from DST Group highlighted that the first of these calls provides critical information. The first call occurred from 18:39:53 to 18:40:56 and is important because the measured BFO is significantly different from the BFO on

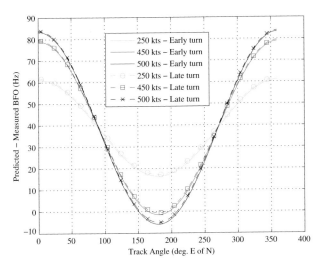

Fig. 5.6 Implied track angles for MH370 with different assumed ground speeds at 18:40, 7 March 2014

the R1200 measurement preceding it at 18:28:15. The R1200 BFO value is consistent with the speed and direction of the aircraft while under radar coverage whereas the later C-channel BFO value is not. Assuming that the change in BFO implies a turn, the difference between the BFO predicted by using a MATLAB model of the SDU software[1] and the measured BFO on the C-channel was analysed as a function of post-turn direction and for a range of aircraft speeds and turn times between 18:28:15 and 18:39:53. Figure 5.6 shows the residual error and it clearly demonstrates that only Southerly track angles are consistent with the C-channel measurements. The model predicted BFO values of Northerly paths are more than 10 standard deviations away from the measured BFO.

The BTO measurements from the R1200 messages at 18:28:15 and 19:41:03 are not consistent with the velocity vector before 18:28:15. The only way to satisfy these measurements and maintain a feasible air speed is for the aircraft to have turned. However, the time window for this turn is more than an hour. The 18:39:53 C-channel measurement is critical because when combined with the 18:28:15 BFO measurement it significantly restricts this turning time window to a little over 11 minutes. Using the C-channel data restricts the aircraft trajectories much more tightly than using only R-channel data.

---

[1]*Note* The difference between the MATLAB model output and the SDU software output was found to be inconsequential to this analysis for determining $\delta f_k^{comp}(\mathbf{x}_k)$ in the SATCOM system model.

5.5 Information Content of Measurements

The information content of the BTO and BFO measurements is illustrated in Fig. 5.7. The figure shows a small part of the likelihood function of the BTO and BFO measurements at 19:41:02. The plots were created assuming an altitude of 30,000 ft. The BFO diagram used an assumed aircraft position of 1°S and 93.6°E and a bias of 150 Hz, which is the tarmac value for the accident flight. The BFO contour shape varies slowly with aircraft position. The figure used the measurement error model described earlier in this chapter, namely zero-mean Gaussian noises with standard deviation 29 μs for BTO, and 7 Hz for BFO.

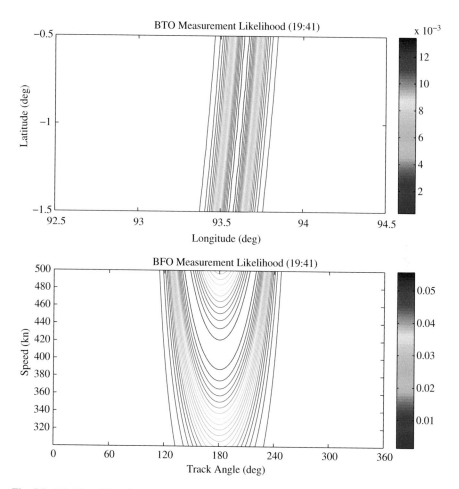

Fig. 5.7 Likelihood functions for BTO and BFO measurements at 19:41:02

The diagrams show that the BTO provides reasonable localisation along a circle of a given range from the satellite. The information provided by the BFO is less precise, providing information on speed, with standard deviation on the order of 50 kn, and direction on the order of $40°$.

Open Access This chapter is distributed under the terms of the Creative Commons Attribution-NonCommercial 4.0 International License (http://creativecommons.org/licenses/by-nc/4.0/), which permits any noncommercial use, duplication, adaptation, distribution and reproduction in any medium or format, as long as you give appropriate credit to the original author(s) and the source, a link is provided to the Creative Commons license and any changes made are indicated.

The images or other third party material in this chapter are included in the work's Creative Commons license, unless indicated otherwise in the credit line; if such material is not included in the work's Creative Commons license and the respective action is not permitted by statutory regulation, users will need to obtain permission from the license holder to duplicate, adapt or reproduce the material.

Chapter 6
Aircraft Cruise Dynamics

The final piece of the Bayesian filter is the dynamics model. In object tracking, it is common to model either the velocity or the acceleration of the object as a random walk in two or three dimensions (e.g., [7]). Such a model has the advantage of mathematical simplicity and can be adequate, particularly in applications where the delay between measurements is low and the area of interest is small. However, an integrated random walk is not adequate for prediction over long time periods, which is required for applications where the delay between measurements is significant. In the case of MH370, the area of interest is vast and the measurements are extremely sparse in time, occurring approximately hourly. The aircraft is believed to have turned at least once after loss of radar contact, so the dynamics model needs to be adaptive enough to accommodate realistic aircraft manoeuvres and yet not so loose as to cause unrealistic spreading of the state posterior pdf over the whole BTO arc. This is not possible with simple integrated random walk models, so more sophisticated approaches are required.

The dynamics model that was adopted for this analysis consists of a random sequence of deliberate manoeuvres interspersed with periods of cruise, in which the speed and direction are almost constant. This model is motivated by [12, 16]. Commercial aircraft typically operate by setting a desired air speed in terms of a Mach number, which is the ratio of the aircraft speed relative to the air mass around it to the speed of sound. During cruise the aircraft autopilot system seeks to maintain a constant Mach number, altitude and control angle (e.g., heading or track), though random fluctuations will occur in each of these.

Three types of manoeuvre were considered: coordinated turns, where the aircraft maintains a constant speed and changes control angle; altitude changes that occur under constant control angle and speed; and accelerations, where the aircraft maintains a constant control angle and altitude and changes speed (Mach number). Details of the statistical models used for manoeuvres are presented in Chap. 7, while details about the cruise model are provided in this chapter.

© Commonwealth of Australia 2016
S. Davey et al., *Bayesian Methods in the Search for MH370*, SpringerBriefs
in Electrical and Computer Engineering, DOI 10.1007/978-981-10-0379-0_6

6.1   Ornstein–Uhlenbeck Process

As discussed above, it would be inappropriate to assume a random walk in either
speed or direction because, over the long duration between measurements, such a
model would either apply significant probability to speeds which are kinematically
infeasible (if the noise strength was significant), or fail to model the statistical variability that flight paths tend to exhibit (if the noise strength was very small). Instead an
Ornstein–Uhlenbeck (OU) process [41] was adopted to model the speed and direction
under control. Under this model, random perturbations are permitted, but the values
tend to revert back to a prescribed set point. Denoting the quantity being controlled
as $x(t)$ and the prescribed set point as $\bar{x}(t)$, the model is specified in continuous
time as

$$dx(t) = -\beta\,[x(t) - \bar{x}(t)]\,dt + d\bar{x}(t) + dv(t). \tag{6.1}$$

The parameters of the model are the reversion rate β; the strength q of the Brownian
motion process $v(t)$; and the set point $\bar{x}(t)$. The set point is accommodated through the
sampling procedure described in Chap. 7, while the reversion rate and noise strength
are selected using a maximum likelihood procedure operating on logged aircraft flight
data from other flights using the same aircraft type, as will be described shortly.

For a linear, time invariant, continuous time stochastic differential equation of the
form

$$dx(t) = Ax(t)dt + dv(t), \tag{6.2}$$

the equivalent discrete process is

$$\mathbf{x}_k = \boldsymbol{\Phi}_k \mathbf{x}_{k-1} + \mathbf{v}_k, \tag{6.3}$$

where $\boldsymbol{\Phi}_k$ is the system matrix, calculated using the matrix exponential

$$\boldsymbol{\Phi}_k = \exp[\mathbf{A}(t_k - t_{k-1})], \tag{6.4}$$

and the covariance of the noise $\mathbf{v}_k$ is

$$\mathbf{Q}_k = \int_{t_{k-1}}^{t_k} \exp[\mathbf{A}(t_k - \tau)]\mathbf{q}(\tau)(\exp[\mathbf{A}(t_k - \tau)])^{\mathsf{T}}d\tau, \tag{6.5}$$

where $\mathbf{q}(\tau)$ is again the strength of the vector Brownian motion [30, 41], here assumed
constant. In this general case $\mathbf{x}$ is a vector and $\boldsymbol{\Phi}$ is a matrix. For the OU process, we
consider a scalar state x and a scalar system matrix Φ

$$x_k = \bar{x}_k + \Phi_k\,(x_{k-1} - \bar{x}_{k-1}) + w_k, \tag{6.6}$$

where $\bar{x}_k = \bar{x}(t_k)$ and

$$\Phi_k = \exp[-\beta(t_k - t_{k-1})]. \tag{6.7}$$

The variance of the process noise w_k is:

$$Q_k = \frac{q}{2\beta}\left(1 - \exp[-2\beta(t_k - t_{k-1})]\right). \tag{6.8}$$

If the difference in time between two samples is large, then the new value of the OU process is only weakly correlated with the previous value. This can be seen by taking the limit of (6.7) and (6.8) as the difference in time increases: the expected value converges to $\bar{x}_k$, and the variance converges to a steady state value of $\frac{q}{2\beta}$.

When an OU process is used to model velocity, the position that results from integrating the velocity is an Integrated Ornstein–Uhlenbeck process. Analytic expressions for the parameters of this process can also be determined in closed form. In the present context, the overall velocity is constructed as a nonlinear composition of several OU processes, as described below. Consequently, position is determined through numerical integration of the velocity over small time steps.

6.1.1 Determining Process Parameters

Logged aircraft flight data from several flights were analysed in order to select the parameters β and q of the OU processes for Mach number, wind, and angle. The log data includes variables such as position, velocity, Mach number, heading, and local wind velocity. In particular the log contains each quantity that is modelled using the OU process, namely Mach number, wind, and direction. This means that a maximum likelihood process can be applied directly (e.g., [37, 44]), rather than requiring more complicated system identification techniques (e.g., [27]).

In the first step, the data was manually segmented into trajectory parts that do not include manoeuvres (such as altitude, speed or direction changes). This yields L sequences, where the lth sequence is $(x_{l,1}, \ldots, x_{l,K_l})$, and the time of element $x_{l,k}$ is $t_{l,k}$. The unknown parameters are the reversion rate β, the noise strength q and the nominal set point $\bar{x}_l$ for each sequence (which is treated as a non-random nuisance parameter). Defining $\boldsymbol{\theta} = (\beta, q, \bar{x}_1, \ldots, \bar{x}_L)$, the overall log likelihood of the observed data is:

$$s(\boldsymbol{\theta}) = \sum_{l=1}^{L}\left\{\log p(x_{k,1}) + \sum_{k=2}^{K_l}\log p_{\boldsymbol{\theta}}(x_{l,k}|x_{l,k-1})\right\} \tag{6.9}$$

$$= C - \frac{1}{2}\sum_{l=1}^{L}\sum_{k=2}^{K_l}\left\{\log(2\pi Q_{l,k}) + \frac{\left[x_{l,k} - \Phi_{l,k}x_{l,k-1} - (1 - \Phi_{l,k})\bar{x}_l\right]^2}{Q_{l,k}}\right\} \tag{6.10}$$

where $\Phi_{l,k}$ and $Q_{l,k}$ are given by (6.7) and (6.8) respectively, and C is a constant that does not depend on the parameter vector $\boldsymbol{\theta}$. Given a value of β, it is possible to maximise over $(q, \bar{x}_1, \ldots, \bar{x}_L)$ in closed form, thus parameter identification can be performed simply using a line search.

In reality, the data stored in the aircraft log files are not exact values, but rather filtered estimates provided by the on-board navigation system. Since the complete information state of the estimators is not provided, the time correlation structure is unknown. However, our interest in the present context is in predictions over significant periods of time, so log data was sub-sampled such that the time between samples ranges between 60 and 300 s. It is assumed that, over this period, the time correlation of the estimates has decayed to such a point that it is dominated by the natural variability of the process.

6.2   Mach Number

In the most common modes of operation, commercial aircraft maintain an approximately constant Mach number, i.e., the ratio between the true air speed of the aircraft, and the speed of sound in the airmass around the aircraft. The speed of sound in an ideal gas is dependent only on the air temperature:

$$c_{sound}(T) = \sqrt{\frac{\gamma \mathcal{R} T}{M}}, \tag{6.11}$$

where $\gamma \approx 1.40$ is the adiabatic index, $\mathcal{R} \approx 8.314 \, \text{J mol}^{-1} \, \text{K}^{-1}$ is the molar gas constant, M is the molar mass of the gas ($\approx 0.02896 \, \text{kg mol}^{-1}$ for dry air), and T is the absolute temperature (in Kelvins) [26].

Thus the true air speed of the aircraft is given by

$$v_{air} = m \times c_{sound}(T) \tag{6.12}$$

where T is the temperature at the aircraft location and m is the Mach number. The consequence of the dependence of speed of sound on temperature is that changes in temperature (including those due to changes in altitude) cause changes in air speed. The air temperature data used in this study was provided by the Australian Bureau of Meteorology, from the Australian Community Climate and Earth-System Simulator (ACCESS) Global model, referred to as ACCESS-G, outlined in [6]. Further details of this data source are discussed in Sect. 6.4.

Even though the aircraft may seek to hold the Mach number constant, fluctuations can and do occur, e.g., due to turbulence and imperfect control. An OU model is used to represent this variability. The parameters $\beta_{mach} = 1.058 \times 10^{-2}$ and $q_{mach} = 2.05 \times 10^{-7} \, \text{s}^{-1}$ were calculated using the procedure in Sect. 6.1.1 based on historical logged flight data, and represent the combination of these sources of fluctuations. The steady state standard deviation under this model is 3.113×10^{-3}. Note that Mach number is dimensionless.

6.2.1 Cost Index

Another commonly used mode of speed control is Cost Index. In this mode, the air speed is selected dynamically to optimise a combination of fuel consumption and travel time [34]. The pilot specifies a Cost Index value, which indicates the relative importance of fuel and travel time. The autopilot system then dynamically selects a Mach number based on the aircraft weight and altitude. This mode was simulated by randomly sampling a Cost Index value and altitude, and then using lookup tables based on proprietary data provided by Boeing.

The effect of the dynamic optimisation is that the air speed will tend to drop over time if the altitude is held constant. The effect is more noticeable at lower altitudes. During normal operation it is typical for the pilot to initiate climbs in altitude and the effect is less noticeable. The effect is not evident in the validation flights studied in Chap. 9 because in each of these flights the aircraft altitude increases with time. For this reason, the majority of the experiments presented in this book do not use the Cost Index mode. Additional experiments were performed using the Cost Index mode for the accident flight. The result was a slight narrowing of the pdf as compared to the Mach number model, as discussed in Sect. 10.7.

6.3 Control Angle

The motion of an aircraft relative to the Earth is influenced by the weather in its environment, particularly the wind. The aircraft heading (i.e., the direction in which the nose points) and air speed determine its motion relative to the local air mass but the motion relative to the Earth is a vector sum of the air velocity and wind velocity

$$\mathbf{v}_{\text{ground}} = \mathbf{v}_{\text{air}} + \mathbf{v}_{\text{wind}}, \tag{6.13}$$

as shown in Fig. 6.1. We refer to the bearing angle of the ground velocity as the aircraft's *track*; elsewhere the term *course* is also used. It is the track that determines motion relative to locations on the Earth's surface, not the direction in which the aircraft's nose points. Similarly, the ground speed is determined from the magnitude of the vector sum in (6.13) and Fig. 6.1. It is generally assumed in this book that the aircraft can drive a control loop to influence Mach number and hence air speed but that the resulting ground speed is somewhat at the mercy of the wind.

The control angle is assumed to follow an OU process as described above for the Mach number. The model parameters for the control angle OU process (in units of radians) are $\beta_{\text{angle}} = 9.792 \times 10^{-3}$, and $q_{\text{angle}} = 4.074 \times 10^{-8}\,\text{rad}^2\,\text{s}^{-1}$.

Between aircraft manoeuvres, the aircraft motion is notionally straight and level. Commercial aircraft are generally designed to operate under autopilot assistance and it would be highly abnormal for a pilot to maintain direct control of this type of aircraft for any extended period. There are five different modes in which the aircraft autopilot can be programmed for steady flight. The first four of these correspond to different definitions of the aircraft control angle $\theta(t)$. The last is used for navigation to a specified location.

The different definitions of control angle enumerate the combinations of maintaining a steady heading or track relative to a magnetic or geographic bearing; these angles are defined in Fig. 6.1. The direction of the Earth's magnetic field is not aligned with the geographic poles and so angles measured with a magnetic compass are not the same as angles measured with respect to the geographic poles. The difference between the two angles is referred to as magnetic declination [11]. Angles measured with respect to the poles are referred to as *true* angles whereas angles measured by compass are referred to as *magnetic* angles. Magnetic declination varies with position on the Earth's surface, so a particular magnetic angle corresponds to different true angles at different locations. Figure 6.2 shows a map of magnetic declination for

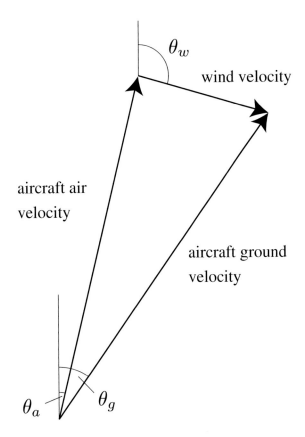

Fig. 6.1 Ground velocity is determined from the vector sum of the velocity relative to the local air mass, and the velocity of the local air mass, i.e., the wind. The *heading* is the direction in which the aircraft is pointing, which is notionally the same as the direction in which it is moving relative to the local air mass, θ_a. The *track* is the direction in which the aircraft is moving with respect to the ground, θ_g

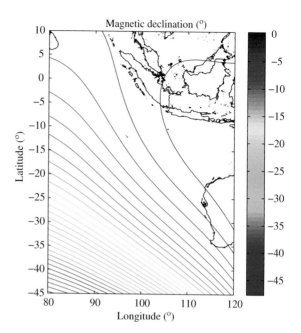

Fig. 6.2 Magnetic declination (degrees) at 40,000 ft on 07 March, 2014. *Source* NOAA [31]

the Indian Ocean region. There are significant variations in magnetic declination in the areas through which the accident flight is thought to have flown. If the aircraft had been following a fixed compass reading then it would appear to gradually turn. Each of the different autopilot modes is now briefly described.

6.3.1 Constant Magnetic Heading

In Constant Magnetic Heading mode, the aircraft aims to hold a steady air speed and magnetic heading, so variations in ground velocity will occur due to variations in wind, and magnetic declination. The bearing of the air velocity is determined by adding the magnetic declination $\phi(\mathbf{x})$ to the control angle

$$\theta_{\text{air}}(t) = \theta(t) + \phi\big(\mathbf{x}(t)\big). \tag{6.14}$$

The North and East ground velocity components follow directly from the vector sum,

$$\mathbf{v}_{\text{ground}}(t) = \begin{bmatrix} \cos\left(\theta_{\text{air}}(t)\right) \\ \sin\left(\theta_{\text{air}}(t)\right) \end{bmatrix} v_{\text{air}}(t) + \mathbf{v}_{\text{wind}}(t). \tag{6.15}$$

6.3.2   Constant True Heading

In Constant True Heading mode, the aircraft aims to hold a steady air velocity,
i.e., air speed and true (non-magnetic) heading. This is the same as Constant Magnetic
Heading except that the magnetic declination is not applied to the control angle,

$$\theta_{\text{air}}(t) = \theta(t). \tag{6.16}$$

6.3.3   Constant Magnetic Track

In Constant Magnetic Track mode, the aircraft aims to hold the direction of the ground
velocity vector, relative to magnetic North, constant. The autopilot automatically
adjusts the aircraft heading to compensate for the wind so that the vector sum of the
air velocity and the wind velocity aligns with the desired track angle. The aircraft
measures the wind, so the direction and magnitude of the wind velocity vector (i.e.
θ_{wind} and v_{wind}) are known. The air speed v_{air} is controlled as is the ground velocity
angle θ_{ground}. Once again the control angle is corrected by the magnetic declination
to give the true track angle

$$\theta_{\text{ground}}(t) = \theta(t) + \phi(\mathbf{x}(t)). \tag{6.17}$$

With reference to the triangle geometry in Fig. 6.1, the lengths of two of the sides
are known and the angle between the track angle and the wind angle defines the
angle shown at the top of the triangle. The ground speed can be determined using
the cosine rule, resulting in:

$$v_{\text{ground}}(t) = \cos\left[\theta_{\text{ground}}(t) - \theta_{\text{wind}}(t)\right] v_{\text{wind}}(t)$$
$$+ \sqrt{v_{\text{air}}(t)^2 - v_{\text{wind}}(t)^2 \sin^2\left[\theta_{\text{ground}}(t) - \theta_{\text{wind}}(t)\right]}. \tag{6.18}$$

6.3.4   Constant True Track

In Constant True Track mode, the aircraft aims to hold the true direction of the ground
velocity vector constant. In terms of the model this is the same as Constant Magnetic
Track except that the magnetic declination is not applied to the control angle,

$$\theta_{\text{ground}}(t) = \theta(t). \tag{6.19}$$

6.3.5 *Lateral Navigation*

The fifth mode is referred to as lateral navigation and implies that the aircraft follows a great elliptical path to a prescribed destination. This is the type of navigation typically used between way points. In this mode a fixed latitude and longitude would be defined and the autopilot would determine the geodesic (shortest constant altitude path) linking this destination with the current location. Under lateral navigation, the aircraft heading and track angles will gradually change. In the present context, lateral navigation is similar to Constant True Track, except that the control angle (i.e., the track) varies as the aircraft moves.

The path travelled under lateral navigation can be determined using Vincenty's reckoning algorithm [46]. The ground velocity is calculated given the local track angle as for Constant True Track, and Vincenty's algorithm is used to calculate both the new location and the updated track angle required to remain on the same great elliptical path (given a small time step size).

Whereas other autopilot modes follow the same control angle indefinitely until the pilot modifies settings, lateral navigation permits the pilot to program a sequence of waypoints that can be either pre-defined, named waypoints, or manually entered coordinates. The autopilot constructs a route which passes through each in turn, following great elliptical paths between them. One way of simulating lateral navigation would be to draw a random sequence of waypoints and this would be quite successful for typical commercial flights. However, custom waypoints can be entered and on an atypical flight could even be likely. In the present context, these deliberate manoeuvres are modelled as changes in the control angle, as with other modes.

Expert advice indicates that if the autopilot system is operating in lateral navigation mode and it reaches the final programmed waypoint, then it reverts to the previously selected heading hold mode, i.e., Constant Magnetic Heading or Constant True Heading. This behaviour has been modelled by simulating a process which switches from lateral navigation to either Constant Magnetic Heading or Constant True Heading at a random time. This is effectively a special manoeuvre that can only happen once. The probability that the switch occurs at time t is modelled as exponentially distributed with a mean time of $6/\ln(2)$ hours, which means that the probability of switching in 6 h is 0.5. The exponential model is discussed in detail in the next chapter.

6.4 Wind

As described in the preceding sections, the aircraft controls air speed and either heading or track angle. In Constant Magnetic/True Heading modes, the wind will influence both the ground speed and the ground track angle. In Constant Magnetic/True Track and lateral navigation modes, the wind will influence the ground speed. For this reason, wind must be incorporated into the analysis.

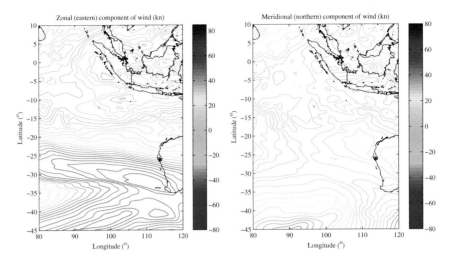

Fig. 6.3 Wind data for 8 March 2014, 00:00, at 175 hPa (approximately 41,000 ft). Contour lines are drawn every 5 kn. The zonal component of the wind is the component blowing toward the East, while the meridional component is the component blowing toward the North

The wind data used in this study was provided by the Australian Bureau of Meteorology from the ACCESS-G model, outlined in [6]. The model provides estimates of wind at air pressures from 150 to 300 hPa in 25 hPa steps and from 300 to 500 hPa in 50 hPa steps. This corresponds to an altitude range from approximately 18,000 to 44,000 ft. The model provides predictions every three hours at latitudes and longitudes spaced by $0.375°$ and $0.5625°$ respectively. Linear interpolation is performed in time, altitude, latitude and longitude. An example of the model output is shown in Fig. 6.3.

The predictions provided by the climate model are averages over time and space, so even if they are very accurate they do not exactly provide the wind experienced at a particular location and time. For this reason, two independent OU processes are used to model the error in the Eastern and Northern components of the wind data. Parameters of the model were determined using the process described in Sect. 6.1.1 based on logged aircraft flight data (i.e., using the predicted wind value as the nominal set point). The OU process (in units of knots) parameter values were found to be $\beta_{wind} = 0.001087$, and $q_{wind} = 0.07021\,\text{kn}^2\,\text{s}^{-1}$. This corresponds to a steady state standard deviation of 5.684 kn.

6.5   Altitude

The measurements available from the Inmarsat satellite are largely insensitive to the aircraft altitude, but modelling of altitude is necessary due to the coupling of altitude, air temperature and air speed (via the Mach number), and due to the variation of wind with altitude. This is modeled by assuming that altitude is constant during cruise, but permitting deliberate changes in altitude as described in Chap. 7.

6.6 Putting It Together

In summary, the process for simulating aircraft cruise dynamics follows these steps:

1. Simulate the Mach number using the OU process:

$$m_k = \bar{m}_k + \Phi_{k,\text{mach}}(m_{k-1} - \bar{m}_{k-1}) + w_{k,\text{mach}}$$

where $w_{k,\text{mach}}$ is a zero-mean, Gaussian random variable with variance $Q_{k,\text{mach}}$, and $\Phi_{k,\text{mach}}$ and $Q_{k,\text{mach}}$ are calculated with (6.7) and (6.8) using β_{mach} and q_{mach}.

2. Simulate the control angle using the OU process:

$$\theta_k = \bar{\theta}_k + \Phi_{k,\text{angle}}(\theta_{k-1} - \bar{\theta}_{k-1}) + w_{k,\text{angle}}$$

where $w_{k,\text{angle}}$ is a zero-mean, Gaussian random variable with variance $Q_{k,\text{angle}}$, and $\Phi_{k,\text{angle}}$ and $Q_{k,\text{angle}}$ are calculated with (6.7) and (6.8) using β_{angle} and q_{angle}.

3. Look up temperature in prediction table, store as T_k. Calculate true air speed as:

$$v_{k,\text{air}} = m_k \sqrt{\frac{\gamma \, \mathcal{R} T_k}{M}}$$

4. Look up wind in prediction table, store values as $\bar{v}_{k,\text{wind}}^{\text{north}}$ and $\bar{v}_{k,\text{wind}}^{\text{east}}$. Simulate the true wind using the OU process:

$$v_{k,\text{wind}}^{\text{north}} = \bar{v}_{k,\text{wind}}^{\text{north}} + \Phi_{k,\text{wind}}(v_{k-1,\text{wind}}^{\text{north}} - \bar{v}_{k-1,\text{wind}}^{\text{north}}) + w_{k,\text{wind}}^{\text{north}}$$
$$v_{k,\text{wind}}^{\text{east}} = \bar{v}_{k,\text{wind}}^{\text{east}} + \Phi_{k,\text{wind}}(v_{k-1,\text{wind}}^{\text{east}} - \bar{v}_{k-1,\text{wind}}^{\text{east}}) + w_{k,\text{wind}}^{\text{east}}$$

where $w_{k,\text{wind}}^{\text{north}}$ and $w_{k,\text{wind}}^{\text{east}}$ are zero-mean, Gaussian random variables with variance $Q_{k,\text{wind}}$, and $\Phi_{k,\text{wind}}$ and $Q_{k,\text{wind}}$ are calculated with (6.7) and (6.8) using β_{wind} and q_{wind}.

5. Calculate ground velocity $\mathbf{v}_{\text{ground}}$ from air speed, control angle and wind velocity based on mode in use, as described in Sect. 6.3.

6. Using ground velocity, predict forward to calculate new position.

The lateral navigation autopilot model uses a time step of 60 s. Although it is possible to determine great elliptical paths with a very high degree of accuracy, the wind is spatially varying and the model samples the residual wind error, so larger steps are not appropriate. The other autopilot models use numerical approximations to propagate the state so a shorter time step is appropriate: we used a time step of 10 s.

Open Access This chapter is distributed under the terms of the Creative Commons Attribution-NonCommercial 4.0 International License (http://creativecommons.org/licenses/by-nc/4.0/), which permits any noncommercial use, duplication, adaptation, distribution and reproduction in any medium or format, as long as you give appropriate credit to the original author(s) and the source, a link is provided to the Creative Commons license and any changes made are indicated.

The images or other third party material in this chapter are included in the work's Creative Commons license, unless indicated otherwise in the credit line; if such material is not included in the work's Creative Commons license and the respective action is not permitted by statutory regulation, users will need to obtain permission from the license holder to duplicate, adapt or reproduce the material.

Chapter 7
Aircraft Manoeuvre Dynamics

As discussed in the previous chapter, the dynamics model used for this analysis consists of a sequence of deliberate manoeuvres interspersed with periods of cruise, in which the speed and control angle are almost constant. This chapter provides details of the statistical models used for manoeuvres. These models describe both the frequency of the aircraft manoeuvres and how the aircraft state changes as a result of each manoeuvre.

Three types of manoeuvre were incorporated into the model: turns, that result in changes to the aircraft control angle; accelerations, that result in changes to the aircraft Mach number; and vertical manoeuvres, that result in changes to the aircraft altitude. Each type of manoeuvre was assumed to occur independently. In practice it is common for multiple changes to occur together. The independent model does not preclude this but it does not favour it either. Without data to build a correlated model, independence is a pragmatic assumption.

7.1 Manoeuvre Frequency

Each of the aircraft manoeuvres can be described by three parameters: the start time of the manoeuvre, the rate of manoeuvre (which will be assumed to be constant), and the extent or duration of the manoeuvre. For example a turn can be defined using its start time, the rate of change of angle, and the angle turned through. We first describe the model for the start time of manoeuvres, which is the same for each manoeuvre type.

The time between manoeuvres was modelled as an exponential distribution. The exponential distribution has a single parameter, which can be interpreted as the average time between events. The different types of manoeuvre are modelled with

© Commonwealth of Australia 2016
S. Davey et al., *Bayesian Methods in the Search for MH370*, SpringerBriefs
in Electrical and Computer Engineering, DOI 10.1007/978-981-10-0379-0_7

a single average manoeuvre period denoted by τ. The use of a single time constant implies an intuitive model where the aircraft can be thought to be either manoeuvring or not, based on the level of pilot interaction. If each manoeuvre was instantaneous, for example the speed changed from Mach 0.83 to Mach 0.81 in zero seconds, then the exponential distribution gives rise to a Poisson arrival process with an average number of T/τ accelerations in T seconds. In practice, the model takes a finite time to change from the pre-manoeuvre state to the post-manoeuvre state and the exponentially distributed delay to the next manoeuvre is applied after the end of the previous manoeuvre. The impact of this is to increase the effective average time between manoeuvres by the average manoeuvre duration. It also imposes a maximum number of manoeuvres in time T whereas the Poisson distribution has infinite support. We assume that each manoeuvre duration is short enough to omit it from the model description below.

The probability of making a turn at time t under the exponential model is given by

$$p(t; \tau) = \frac{1}{\tau} \exp\left\{-\frac{t}{\tau}\right\}, \tag{7.1}$$

and the probability of having no turns in the interval $[0, T]$ is

$$p(t > T; \tau) = \exp\left\{-\frac{T}{\tau}\right\}, \tag{7.2}$$

Using this model, and neglecting the turn duration, the prior likelihood of a sequence of N turns at times $0 \le t_1 \le \ldots t_N \le T$ is

$$\begin{aligned}
p(t_{1:N}; \tau) &= \left[\prod_{n=1}^{N} p\left(t_n | t_{1:n-1}; \tau\right)\right] p\left(t_{N+1} > T | t_{1:N}\right) \\
&= \left[\prod_{n=1}^{N} \frac{1}{\tau} \exp\left\{-\frac{t_n - t_{n-1}}{\tau}\right\}\right] \exp\left\{-\frac{T - t_N}{\tau}\right\} \\
&= \tau^{-N} \exp\left\{-\frac{T}{\tau}\right\},
\end{aligned} \tag{7.3}$$

where $t_0 = 0$ is not a turn time but makes the notation more convenient.

Similar expressions hold for the accelerations and vertical manoeuvres. Since all three manoeuvres are assumed to occur independently, the probability of a sequence of turns, accelerations and vertical manoeuvres is the product of the turn sequence probability, the acceleration sequence probability and the vertical manoeuvre sequence probability.

7.2 Manoeuvre Extent

The model requires a statistical description of how the aircraft manoeuvres as well as when it manoeuvres. As stated above, we assume that all manoeuvres happen at a constant rate; for example, the heading angle could change steadily at $0.5° \, s^{-1}$. For real aircraft manoeuvres, the precise value of this rate can be different but in the context of the flight prediction required for MH370, it is possible to assume fixed values without losing diversity of the sampled paths. Different rates of turn and longitudinal accelerations were explored in earlier versions of the model but these changes simply resulted in a negligible increase in position uncertainty after the manoeuvre.

The aircraft turns were assumed to follow a simple bank model where the relationship between the bank angle and the angular velocity of a turn follows

$$\omega = \frac{\tan(\theta_{bank}) \, g}{v}, \tag{7.4}$$

where θ_{bank} is the aircraft bank angle and g is the acceleration due to gravity. Tight turns with a high angular velocity require low speed or high bank angle. Varying the bank angle was found to produce only minor variations in the overall aircraft trajectory since turns were only a small portion of the trajectories, so a fixed bank angle of $15°$ was chosen. This is nominally a steep bank for a commercial aircraft, although well within its performance limits. At an air speed of 500 kn, for example, the turn rate would be approximately $0.6° \, s^{-1}$ and a $90°$ turn would take 2.5 min to perform.

The final parameter required to characterise a turn is the duration of the turn, or equivalently the total change in angle. The aircraft was assumed to not make turns of more than $180°$ and the change in angle was uniformly sampled from $-180°$ to $180°$. It is possible for the model to sample two or more turns in very quick succession and as such looping turns of more than $180°$ are possible, though unlikely. The likelihood of a quick succession of turns like this depends on the mean time τ.

The accelerations were also assumed to occur with a fixed rate of change of Mach. The assumed rate corresponds to a change of Mach of 0.1 in one minute. This nominal number was not matched to aircraft performance standards but instead simply acts to define a finite duration for each speed change. The Mach number after an acceleration was uniformly sampled from 0.73 to 0.84; this range was chosen on the basis of expert advice to ensure that the required flight endurance is achievable.

The vertical manoeuvres were assumed to occur at a nominal rate of 4,000 ft per minute. Again this rate is not meant to model the actual typical behaviour of commercial aircraft. Rather it provides a reasonable finite time extent to the manoeuvre. The altitude after manoeuvre was uniformly selected in steps of 1,000 ft between 25,000 and 43,000 ft.

Using finite time to execute manoeuvres rather than making instantaneous changes allows for the possibility that a measurement could have been collected during a manoeuvre. The model retains the ground velocity vector but not the altitude rate, so in principle it could describe measurements collected part way through a turn or acceleration but not during climbs or descents. The vertical rate will affect a BFO measurement but since the model only uses a nominal vertical rate it is unlikely to match any actual vertical manoeuvre in detail. In practice, if a measurement was collected during a turn or acceleration it would be very difficult for the filter to infer the trajectory since it would need to model the instantaneous ground speed at the measurement time to match the measured BFO. For the C-channel data at 18:39 and 23:15 there are clusters of BFO measurements and these appear to be statistically stationary, i.e., they do not support the premise that the aircraft is turning during collection.

The three types of manoeuvre are independent and the distribution of manoeuvre extents is uniform for each type, the overall prior probability of a sequence of turns $[(t_{\theta,1}, \theta_1), \ldots, (t_{\theta,N_\theta}, \theta_{N_\theta})]$, accelerations $[(t_{a,1}, v_1), \ldots, (t_{a,N_a}, v_{N_a})]$ and altitude changes $[(t_{h,1}, h_1), \ldots, (t_{h,N_h}, h_{N_h})]$ is given by

$$p\left(\{t_{\theta,1:N_\theta}, \theta_{1:N_\theta}\}, \{t_{a,1:N_a}, v_{1:N_a}\}, \{t_{h,1:N_h}, h_{1:N_h}\}; \tau\right)$$
$$= \exp\left\{-\frac{3T}{\tau}\right\}\left(\tau\delta\theta\right)^{-N_\theta}\left(\tau\delta a\right)^{-N_a}\left(\tau\delta h\right)^{-N_h}, \qquad (7.5)$$

where $\delta\theta = 2\pi$ rad is the span of the uniform turn angle distribution, $\delta a = 0.11$ (Mach) is the span of the uniform speed distribution, and $\delta h = 19$ is the number of possible discrete altitude values.

7.2.1 Parameter Selection

The exponential delay model is parameterised by the average time between events, τ. It is not clear what this value should be or even if it should be constant. In particular one could argue that the behaviour of the aircraft during the accident flight did not match typical commercial aircraft. The average delay is therefore treated as a constant hyperparameter that is potentially different for every hypothesised flight trajectory, but fixed over time.[1] A Jeffreys prior [24] was applied to the time constant; this prior is a non-informative distribution in the sense that it is invariant to the parameterisation used. For example, it yields the same result for the two parameterisations of the exponential distribution, i.e., the rate parameterisation and

---

[1] Additional experiments were conducted in which the parameter was permitted to change at 19:41 (well after the initial manoeuvre), but little change to the final distribution was evident.

the time-constant parameterisation. The Jeffreys prior is proportional to the square root of the determinant of the Fisher information; in the case of an exponential distribution, it is given by

$$p(\tau) = \begin{cases} K(\tau_1, \tau_2)\, \tau^{-1} & \tau_1 \leq \tau \leq \tau_2 \\ 0 & \text{otherwise} \end{cases}, \tag{7.6}$$

where $K(\tau_1, \tau_2) = \left(\log(\tau_2) - \log(\tau_1)\right)^{-1}$. A support range of $0.1 < \tau < 10\,\text{h}$ between events was chosen for each manoeuvre type, spanning the range of cases when manoeuvres occur every few minutes, to when manoeuvres are rare in the entire flight. A single, common parameter was drawn for all three types of manoeuvre.

The joint pdf of the full manoeuvre description is given by

$$p\left(\{t_{\theta,1:N_\theta}, \theta_{1:N_\theta}\}, \{t_{a,1:N_a}, v_{1:N_a}\}, \{t_{h,1:N_h}, h_{1:N_h}\}, \tau\right)$$
$$= p(\tau)\exp\left\{-\frac{3T}{\tau}\right\} \left(\tau \delta \theta\right)^{-N_\theta} \left(\tau \delta v\right)^{-N_a} \left(\tau \delta h\right)^{-N_h}. \tag{7.7}$$

7.2.2 Manoeuvre Model Summary

Figure 7.1 shows a graphical representation of the aircraft dynamics model. Ignoring measurements (which are not illustrated), the sequence of manoeuvre times and extents are conditionally independent from other kinds of manoeuvres given the

Fig. 7.1 Graphical model showing the conditional dependencies in the probabilistic model, where $\mathbf{X} = (\mathbf{x}_0, \ldots, \mathbf{x}_K)$ represents the entire trajectory

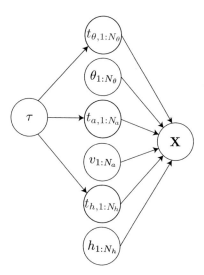

common manoeuvre time constant. The aircraft state forms a Markov chain conditioned on the manoeuvre parameters. The measurements (not illustrated) are conditionally independent from each other given the state sequence.

7.3 Example Realisations

In order to illustrate the behaviour of the complete dynamics model, Fig. 7.2 shows example realisations of trajectories randomly sampled from the model. We emphasise that these do not incorporate information from the SATCOM measurements, but rather represent samples of the prior distribution over trajectories (e.g., $p(\mathbf{x})$ in (3.3)). Each trajectory starts with independent samples of the manoeuvre time constant and initial location and velocity sampled from the accident flight prior, and then draws an independent set of sampled turns, and speed and altitude changes. The prior starts the particles at 18:02 and the figure shows each trajectory through to 00:19, the time of the final SATCOM message. As the figure shows, some of the trajectories exhibit many manoeuvres, while others do not turn or change speed at all.

The histogram of the number of turns and speed changes hypothesised by the prior model is shown in Fig. 7.3. The model is in effect a mixture of Poisson distributions, consisting of a continuum of components ranging from very low times between manoeuvre (six minutes) to very high times between manoeuvre (ten hours). The distribution is the same for each type of manoeuvre since the prior distributions of time constants are the same.

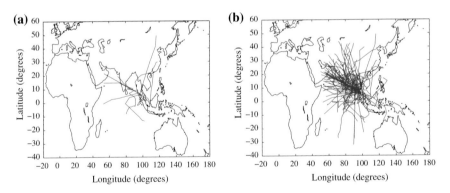

Fig. 7.2 Example trajectories sampled from prior distribution of flight dynamics: **a** 10 random realisations; **b** 100 random realisations

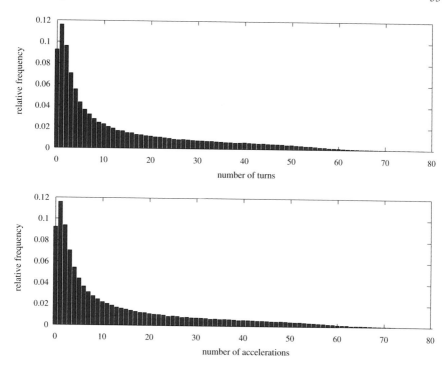

Fig. 7.3 Histograms of the number of manoeuvres selected by 200,000 independent realisations of the model (i.e., not using any BTO/BFO measurements)

Although it is unlikely that any given trajectory will travel for more than six hours with no manoeuvre, those that do all end up in roughly the same place. Figure 7.4 shows a contour representation of the pdf of latitude and longitude at 00:19 that was generated using 200,000 particle draws from the prior model with no measurements and an isotropic kernel. The red diamond in the figure shows the mean of the prior. There is a clear peak corresponding to paths that make no manoeuvre. This is smeared around an arc corresponding to paths that turn once but maintain a steady speed and also radially from the start point corresponding to paths that do not turn but do change speed. The model samples a hyperparameter for the time between manoeuvres from a prior that covers 0.1 to 10 h. When a particle samples a very high value for the time constant it is unlikely to choose to manoeuvre. Where manoeuvres were selected it was again unlikely to choose to change speed and direction, leading to the contours shown. The altitude behaviour of the model is not visualised in the figure. There is an approximately circular region around the initialisation point where the pdf is elevated. This corresponds to particles that have sampled a very small time between manoeuvres: these turn so much that they circle back on themselves repeatedly and are effectively trapped close to their starting location. Beyond this circle, the pdf is

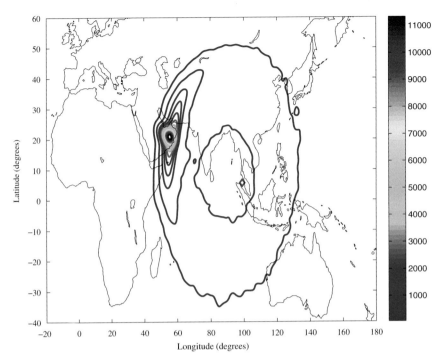

Fig. 7.4 Distribution of position at 00:19 using only model predictions (i.e., not using any of the BTO/BFO measurements). The *red* diamond marks the mean on the prior

lower behind the start point than in front of it. It is also clear that the search region is in a very low probability part of the pdf: this indicates that the prior is not biased towards selecting a particular part of the 00:19 arc. The search region is inside the lowest probability contour used in the map.

Open Access   This chapter is distributed under the terms of the Creative Commons Attribution-NonCommercial 4.0 International License (http://creativecommons.org/licenses/by-nc/4.0/), which permits any noncommercial use, duplication, adaptation, distribution and reproduction in any medium or format, as long as you give appropriate credit to the original author(s) and the source, a link is provided to the Creative Commons license and any changes made are indicated.

The images or other third party material in this chapter are included in the work's Creative Commons license, unless indicated otherwise in the credit line; if such material is not included in the work's Creative Commons license and the respective action is not permitted by statutory regulation, users will need to obtain permission from the license holder to duplicate, adapt or reproduce the material.

Chapter 8
Particle Filter Implementation

Solution of the Bayesian estimation method described in Chap. 3 requires one to recursively integrate the aircraft dynamics pdf (3.6) and multiply it by the likelihood (3.7). Since the measurement model is highly nonlinear and the dynamics model is hybrid discrete-continuous, there is no way to produce a closed form posterior distribution. An alternative is to approximate the distribution numerically. As introduced in Sect. 3.2, the Sample-Importance-Resample (SIR) particle filter draws random samples from the dynamics model and weights them according to the measurement likelihood. This amounts to approximating the posterior distribution as

$$p(\mathbf{x}_k|\mathbf{Z}_k) \approx \sum_{p=1}^{P} w_k^p \delta\left(\mathbf{x}_k - \mathbf{x}_k^p\right),\tag{8.1}$$

where the w_k^p are referred to as weights (and sum to unity) and the $\mathbf{x}_k^p$ are referred to as particles. The convergence properties of this approximation in the limit as the number of particles P increases have been well studied, e.g. [14, 21]. In the SIR version of the particle filter, the particles are randomly generated from the dynamics model and the weights are

$$w_k^p \propto \prod_{k':t_{k'} \leq t_k} p\left(\mathbf{z}_{k'}|\mathbf{x}_{k'}^p\right).\tag{8.2}$$

A problem with sampling from the dynamics is that this can be a very diffuse distribution. In the MH370 case, the model allows for turns and speed and altitude changes, and potentially several of each can be sampled between measurements. The proportion of particles that sample a trajectory close to the measurements will be small and a very large number of samples will be required to capture the high probability regions. This is a well known issue for filtering in high dimensional state spaces.

© Commonwealth of Australia 2016
S. Davey et al., *Bayesian Methods in the Search for MH370*, SpringerBriefs in Electrical and Computer Engineering, DOI 10.1007/978-981-10-0379-0_8

Resampling is one strategy that is used to improve the number of particles follow-ing trajectories with relatively high likelihood. It does this in a sequential manner, in which at each time step unlikely particles are replaced with copies of high likely particles though a random sampling process. Initial approaches used these conven-tional techniques, but it was found to be preferable to be able to process very high particle counts and to adaptively increase the number of particles used until it was possible to identify an adequate number of likely paths, rather than processing a pre-specified number of particles for each time step sequentially. To achieve this par-ticles were propagated and weighted individually; this also reduced the size of the data structures required and allowed preliminary results to be extracted as the filter was executing. The approach adopted was a form of branching mechanism which repeatedly constructs full trajectories.

The method resampled each particle separately, branching a new set of particles from each parent instead of resampling a fixed number of particles across all of the particles at a given time. The branching naturally leads to an exponential growth in the number of particles with time and this was mitigated by pruning extremely unlikely paths when their likelihood became too low. This approach is not necessarily computationally efficient, but in this particular application it was more important to broadly explore the enormous state space than to minimise computational effort.

To motivate the approach, suppose that the distribution prior to resampling is approximated by (8.1), such that integrals can be approximated as

$$\int f(\mathbf{x}_k)p(\mathbf{x}_k)d\mathbf{x}_k \approx \sum_{p=1}^{P} w_k^p f\left(\mathbf{x}_k^p\right), \tag{8.3}$$

where it is assumed that $\sum_{p=1}^{P} w_k^p = 1$. For each particle p, draw $n_k^p \geq 0$ copies of that particle, where n_k^p is a random variable, and for each new particle $\tilde{\mathbf{x}}_k^{\tilde{p}} = \mathbf{x}_k^p$. To each we apply the weight

$$\tilde{w}_k^{\tilde{p}} = \frac{w_k^p}{\mathbb{E}[n_k^p]}. \tag{8.4}$$

Assuming that $\tilde{p}$ indexes the full set of $\tilde{P} = \sum_{p=1}^{P} n_k^p$ new particles, it can easily be shown that:

$$\mathbb{E}\left[\sum_{\tilde{p}=1}^{\tilde{P}} \tilde{w}_k^{\tilde{p}} f\left(\tilde{\mathbf{x}}_k^{\tilde{p}}\right)\right] = \sum_{p=1}^{P} \mathbb{E}[n_k^p]\tilde{w}_k^p f\left(\mathbf{x}_k^p\right) = \sum_{p=1}^{P} w_k^p f\left(\mathbf{x}_k^p\right) \tag{8.5}$$

Thus resampling can also be implemented through a randomised branching proce-dure, recursively adapting the number of particles. This permits a form of depth-first search, which adaptively performs more branching when likely paths result, and tends to prune paths which have low probability.

For our experiments, we chose a procedure which branches quite aggressively when likely paths are discovered, and prunes extremely unlikely paths. Likely paths are duplicated to form $\bar{n}$ branches. This is implemented by setting

$$
p(n_k^p) = \begin{cases} \delta(n_k^p - \bar{n}), & w_k^p \geq \eta \\ w_k^p \delta(n_k^p - 1) + (1 - w_k^p)\delta(n_k^p), & \text{otherwise} \end{cases} \tag{8.6}
$$

for $\eta \ll 1$. Thus, for particles $\tilde{p}$ sampled from parent particle p with $w_k^p \geq \eta$, $\tilde{w}_k^{\tilde{p}} = w_k^p/\bar{n}$, and wide branching will occur, while for particles $\tilde{p}$ sampled from parent particle p with $w_k^p < \eta$, $\tilde{w}_k^{\tilde{p}} = 1$, but most commonly the sub-tree will be pruned.

8.1 BFO Bias

The BFO measurement has a bias term that was not able to be adequately calibrated, as discussed in Sect. 5.3. The model treats this bias as an random variable with a given prior density. It is possible to sample the bias along with the aircraft states but a more efficient implementation is to use a Rao-Blackwellised particle filter [15, 29, 38]. Conditioned on the other states we can write a simplified BFO measurement model

$$
z_k^{\text{BFO}} = \hat{z}_k^{\text{BFO}} + \delta f^{\text{bias}} + w_k^{\text{BFO}}, \tag{8.7}
$$

where $\hat{z}_k^{\text{BFO}}$ is constant because all of the other states, such as aircraft location and velocity, are known because of the conditioning. This conditional measurement equation is clearly linear in the bias and the noise is modelled as Gaussian, so the posterior distribution of the bias can be determined using a Kalman filter update.

8.2 Algorithm

In practical terms, the algorithm proceeds by repeating the following process for each particle:

1. Randomly sample an average time between manoeuvres τ.
2. Randomly sample a starting state $\mathbf{x}_0$ (position, Mach, control angle and altitude) from the prior at $t_0 = 18{:}01{:}49$, which is described in Chap. 4.
3. Initialise the BFO bias Kalman filter.
4. Perform the following recursion, starting with the sample at $\mathbf{x}_0$ and measurement time index $k = 1$:

    a. Draw a sample of the trajectory from $\mathbf{x}_{k-1}$ to $\mathbf{x}_k$ using the hyperparameter τ for selection of turns, speed changes and altitude changes.

b. Calculate the measurement likelihood $p(\mathbf{z}_k|\mathbf{x}_k)$ and use it to update the trajectory weight $w_k^p = p(\mathbf{z}_k|\mathbf{x}_k) w_{k-1}^p$.

c. Use the sampled trajectory to update the BFO bias Kalman filter.

d. If we have reached the final measurement $k = K$, store the trajectory and weight.

e. Otherwise, if the accumulated weight is too low, i.e., $w_k^p < \eta$ then branch a single time with probability w_k^p and weight $\tilde{w}_k^{\tilde{p}} = 1$, terminating the recursion branch with probability $(1 - w_k^p)$; otherwise, branch $\bar{n}$ times to process remaining time steps with weight $\tilde{w}_k^{\tilde{p}} = w_k^p/\bar{n}$.

The particle weights constructed by the method are not normalised. A normalisation step is performed when the final set of weights at the last time point is used to construct the required pdf. The process in step 4a, namely sampling a trajectory, is critical and is realised through a finite time difference implementation given by the following steps:

1. Randomly sample times to make the next turn, speed change and altitude change
2. While the current sample time is before the next measurement time, t_k

Table 8.1 State vector elements

State vector	
Latitude	Degrees
Longitude	Degrees
Mach number set point	The OU parameter, denoted $\bar{m}_k$ in Sect. 6.6
Instanteous Mach number	Mach, denoted m_k in Sect. 6.6
Control angle set point	The OU parameter, denoted $\bar{\theta}_k$ in Sect. 6.6
Instanteous control angle	Control angle, denoted θ_k in Sect. 6.6
Altitude	In discrete 1,000 feet steps
Instantaneous wind error, north	The difference between the wind and the tabulated value, denoted $v_{k,\text{wind}}^{\text{north}}$ in Sect. 6.6
Instantaneous wind error, east	The difference between the wind and the tabulated value, denoted $v_{k,\text{wind}}^{\text{east}}$ in Sect. 6.6
BFO bias mean	Mean of the Kalman filter used to estimate the BFO bias δf^{bias}
BFO bias variance	Variance of the Kalman filter used to estimate the BFO bias δf^{bias}
Mean time to manoeuvre	τ
Autopilot control angle mode	Used to choose between constant true/magnetic heading/track and lateral navigation
Cost index value, when used	An integer between 0 and 100

a. If the current sample time is the time of a manoeuvre (turn, speed change or altitude change), then execute the manoeuvre and sample a new time to make the next manoeuvre

b. Otherwise predict ahead 10 s (or to the next manoeuvre or measurement, whichever occurs first)

A manoeuvre is executed by making a sequence of 1 s steps. For each step the angle, speed or altitude is incremented and the aircraft position is predicted ahead. The increments continue until the new desired angle, speed or altitude is achieved. The procedure for state prediction under cruise dynamics is summarised in Sect. 6.6. The state vector used for the model is given in Table 8.1. There are a large number of parameters involved with this model and the full description of these is provided in Table 8.2.

Table 8.2 Summary of filter parameters

Initialisation		
Latitude	Gaussian	s.d. 0.4 min
Longitude	Gaussian	s.d. 0.4 min
Control Mach	Gaussian	s.d. 0.03
Control angle	Gaussian	s.d. 1°
BFO bias	Gaussian	s.d. 25 Hz
Mach deviation	Gaussian	s.d. 0.00311
Angle deviation	Gaussian	s.d. 0.0826°
North wind deviation	Gaussian	s.d. 5.68 kn
East wind deviation	Gaussian	s.d. 5.68 kn
Cruise		
Mach	Reversion rate β_{mach}	1.06×10^{-2}
	Noise strength q_{mach}	2.05×10^{-7} s^{-1}
Control angle	Reversion rate β_{angle}	9.8×10^{-3}
	Noise strength q_{angle}	4.07×10^{-8} rad^2 s^{-1}
North/East wind	Reversion rate β_{wind}	1.09×10^{-3}
	Noise strength q_{wind}	7.02×10^{-2} kn^{2}s^{-1}
Manoeuvres		
All	Mean manoeuvre time τ	$\sim$Jeffreys(0.1, 10)
Mach	Uniform new Mach	0.73–0.84
Control angle	Uniform turn angle	$\pm 180°$
Altitude	Uniform new altitude	25,000 to 43,000 ft
Implementation		
Branching rate	Constant $\bar{n}$ within a flight	3–10
Likelihood threshold	Constant η within a flight	e^{-25} or e^{-30}

8.3 Assumptions

The key assumptions used by the filter are:

1. The radar data provides an accurate estimate of the aircraft trajectory up to 18:01:49. If, for example, the radar track used to build the prior were actually from a different aircraft, the predicted pdf would be invalid. Discarding the radar data leads to a significant broadening of the search zone, and accident investigators believe the radar data to be correctly associated with MH370. Chapter 10.6 considers an alternative analysis which ignores the radar data.
2. The measurement error characteristics are known. The pdf of BTO and BFO measurements, in particular the standard deviation of each, is provided to the algorithm as a known input. Extensive study of the statistics of these measurements has been undertaken and the distributions assumed are well characterised, subject to the caveats discussed in Chap. 5. Incremental changes, such as minor inflation of the assumed BTO variance would lead to incremental changes in the filter output.
3. The aircraft cruises in one of five prescribed modes and does not change between them (other than a single possible change from lateral navigation to constant magnetic/true heading). It is possible that the whole flight was continually under manual control but it is highly unlikely. The use of typical autopilot modes is reasonable.
4. Infinite fuel: the fuel constraints on the aircraft can be applied to the pdf afterwards. In the simplest case, maximum reachable ranges could be used to censor impossible trajectories. However, analysis of candidate trajectories has indicated that the majority are feasible. Broad information about the fuel consumption rate of the aircraft has been used to inform the range of allowable Mach numbers.
5. The fluctuations in speed, angle and the error in wind velocity are well-modelled by the OU process. The parameters of the OU model were selected to model these quantities based on recorded data from real flights.
6. The random turn and speed change model is rich enough to describe the real aircraft dynamics and the implicit preferred path for the model does not bias prediction. Validation results in the next chapter show that the model successfully produces pdfs containing the true aircraft location for the available instrumented flights that include air speed changes, altitude changes and angle changes.
7. The aircraft air speed is limited to the range Mach 0.73–0.84. Fuel consumption becomes very inefficient at speeds higher than this and at lower speeds the aircraft is not able to match the measurements. In practice it is likely that the viable range of speeds is actually much narrower than this.
8. In Chap. 10, the pdf of the location of the aircraft at 00:19 is combined with a distribution of aircraft translation during descent, to give a final search zone.

This distribution was developed by ATSB [5] and largely determines the width of the search area along the 00:19 arc. It is assumed that this distribution adequately models the true descent scenario.

Open Access This chapter is distributed under the terms of the Creative Commons Attribution-NonCommercial 4.0 International License (http://creativecommons.org/licenses/by-nc/4.0/), which permits any noncommercial use, duplication, adaptation, distribution and reproduction in any medium or format, as long as you give appropriate credit to the original author(s) and the source, a link is provided to the Creative Commons license and any changes made are indicated.

The images or other third party material in this chapter are included in the work's Creative Commons license, unless indicated otherwise in the credit line; if such material is not included in the work's Creative Commons license and the respective action is not permitted by statutory regulation, users will need to obtain permission from the license holder to duplicate, adapt or reproduce the material.

Chapter 9
Validation Experiments

The variable rate model developed for MH370 was validated by analysing data from a collection of flights where the true aircraft location was known; we refer to these as validation flights. A total of six validation flights were used for testing. Data was available from a larger number of flights but the majority of these were in relatively short segments of less than three hours. There were only a few that maintained communications with the satellite Inmarsat-3F1 for longer periods and it was not thought productive to examine the prediction performance over time segments shorter than three hours. Of the six flights, four are previous flights of the accident aircraft, 9M-MRO, and the other two are flights of different aircraft that occurred at the same time as the accident flight. Three of the flights are relatively short and are between locations inside Asia, and the other three are flights from Asia to Europe.

The data available for the accident flight consists of mostly R1200 communication messages at approximately one hour intervals. In order to emulate the measurement information content, measurement data sets were formed by randomly sub-sampling R1200 communication messages from the validation flights. Ten different subsets were formed for each validation flight, resulting in a total of sixty validation measurement sets. Multiple sets were drawn from each flight to increase the statistical significance of the testing data set. They also serve to illustrate the sensitivity of the method to the precise measurement times and values. The measurement subsets were selected using a randomised process that aimed to achieve an average time between measurements of one hour. For the analysis we treat the measurement subsets as independent Monte Carlo trials. However there are several variables that are in common within the group of ten subsets of a single flight: the aircraft geometry is obviously the same for each subset since they are drawn from the same flight; the residual wind errors are the same; and the BFO is known to have a slowly varying bias, so there can be correlation in the BFO measurements from different subsets if those subsets choose measurements at similar times. Finally, some subsets may in fact randomly choose the same measurement as another subset.

© Commonwealth of Australia 2016
S. Davey et al., *Bayesian Methods in the Search for MH370*, SpringerBriefs
in Electrical and Computer Engineering, DOI 10.1007/978-981-10-0379-0_9

In each validation flight, the true aircraft location was obtained from the Aircraft Communications Addressing and Reporting System (ACARS) data logs. Sections of the flight immediately after take-off and prior to landing were not included in the analysis since the aircraft dynamics are very different at these times and it is unlikely that sparse satellite messages would be sufficient to follow it. For the longer flights into Europe, the aircraft changed satellites part way through the flight so it was not possible to use the whole flight: these were truncated near the end of messaging via the Indian Ocean Region satellite. The filter was initialised using the true aircraft location, speed and control angle with a Gaussian random error. The standard deviation of the initialisation error was chosen to be the same as the prior for the accident flight, that is 0.4° in latitude and longitude, 1° in angle and Mach 0.03 in air speed. For every subset the posterior pdf at the final measurement was predicted ahead to a common time, corresponding to an exact ACARS reporting time. This predicted pdf is compared with the ACARS report.

This chapter first explains the particular characteristics of each flight and presents an example output pdf for one of the measurement sets. This output is subjectively compared with the ACARS truth. The statistical analysis is then presented using an objective performance measure over the sixty validation subsets. Table 9.1 lists the six validation flights used for the analysis and gives comments on some of the characteristics of each. The flights are ordered by time.

Table 9.1 Summary of validation flights

Flight path	Date	Duration (h:mm)	9M-MRO	Comments
Kuala Lumpur to Amsterdam	26 February	7:35	Yes	Eclipse
Mumbai to Kuala Lumpur	2 March	3:20	Yes	Short and almost straight with a gradual late veer, outlier BFO measurements
Kuala Lumpur to Beijing	6 March	4:25	Yes	Single climb, several S-turns
Beijing to Kuala Lumpur	7 March	4:55	Yes	Significant climbs, Mach changes and turns, contains anomalous BTO measurements
Kuala Lumpur to Amsterdam	7 March	7:50	No	Large S-turns
Kuala Lumpur to Frankfurt	7 March	7:03	No	Mid flight heading deviations, outlier BFO measurements

9.1 9M-MRO 26 February 2014 Kuala Lumpur to Amsterdam

The first validation flight was from Kuala Lumpur to Amsterdam on 26 February 2014. This flight was around 7.5 h long but is relatively straight. Figure 9.1 summarises the features of the flight: the upper plot shows a geographic plan; the lower three plots show the aircraft altitude as a function of time, the aircraft heading as a function of time, and the aircraft Mach number as a function of time. Vertical dotted lines show the start and end of the time segment selected for the test. This flight contained an eclipse event so the validation also supports the Inmarsat eclipse correction [2].

Figure 9.2 shows the filtered pdf for the Kuala Lumpur to Amsterdam flight visualised using a three dimensional representation in Google Earth. The filter pdf is defined over a high dimensional space but for visualisation we examine the marginal position distribution in latitude and longitude. Because the BTO measurement error is relatively small the position distribution is centred on an arc of zero BTO error and has a narrow off-arc width. For the visualisation we marginalise the distribution onto the zero BTO error arc and encode the probability density for each point along the arc using altitude: points on the curve higher above the earth correspond to higher probability. A white curve on the map marks the ACARS reported aircraft location, a yellow marker denotes the location of the aircraft at each measurement time. The figure also shows a representative selection of the paths sampled by the filter. The selection shows the highest probability path arriving at each point around the arc: the colour of the path shows the marginal probability at that location on the arc (using a colour map similar to Fig. 5.7, i.e., blue is least likely, red is most likely).

There are a number of paths that end in significantly different locations to the truth. These occur because in this flight the aircraft travels in a direction that is almost horizontally radial from the satellite. While the aircraft moves towards the satellite its initial dynamics constrain the plausible paths but once it passes through the point of closest approach and begins to move away then it is possible to make turns that result in different near-radial paths. The support of these ambiguous paths is disjoint because of the finite number of samples: the true underlying pdf has support all the way around the arc. Without dynamic constraints the location of the peak of the pdf is simply a function of measurement noise.

9.2 9M-MRO 2 March 2014 Mumbai to Kuala Lumpur

The flight from Mumbai to Kuala Lumpur is the shortest validation flight selected. Figure 9.3 summarises the features of the flight: there is a single minor altitude change and the Mach number remains relatively constant. The aircraft heading gradually reduces for most of the flight, turning the aircraft more to the North but a veer near the end turns it back to the South-East. The BFO measurements for this flight contain several outliers that are more than 30 Hz away from other measurements at similar times.

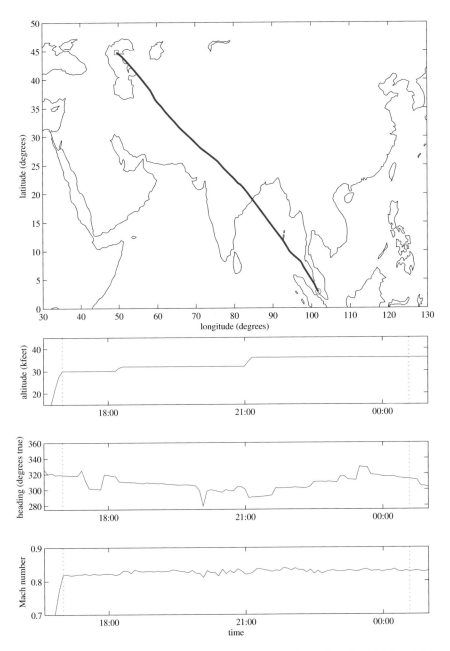

Fig. 9.1 Validation flight 26 February 2014 Kuala Lumpur to Amsterdam. *Vertical dotted lines* show the start and end times of the segment used for validation

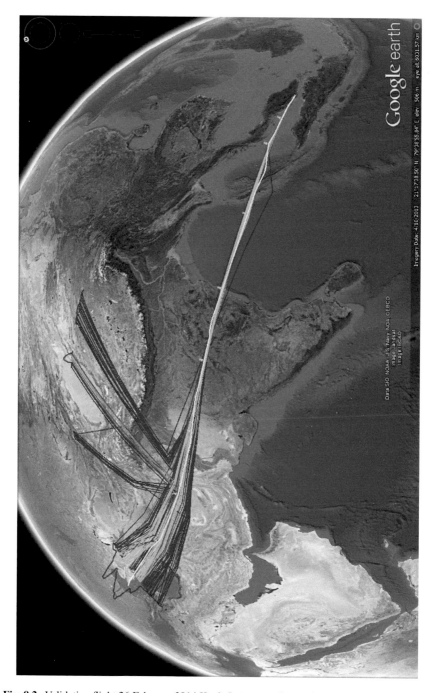

Fig. 9.2 Validation flight 26 February 2014 Kuala Lumpur to Amsterdam

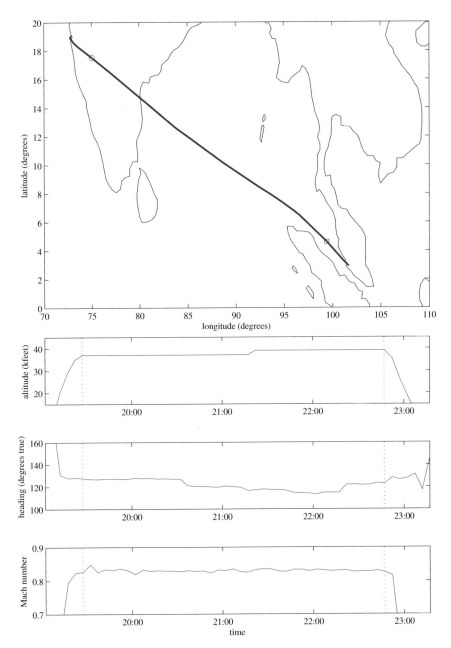

Fig. 9.3 Validation flight 2 March 2014 Mumbai to Kuala Lumpur. *Vertical dotted lines* show the start and end times of the segment used for validation

Figure 9.4 shows the pdf output from the filter; the true ACARS aircraft location is again under the main peak of the pdf. The pdf might appear to be relatively spread compared with some of the other flights, but the scale is much smaller in this case because the flight is short.

9.3 9M-MRO 6 March 2014 Kuala Lumpur to Beijing

This flight is the MH370 route from Kuala Lumpur to Beijing that was flown by the accident aircraft 9M-MRO on 6 March 2014, i.e., the day prior to the accident flight. Figure 9.5 summarises the features of the flight: the flight contained a single altitude change and several turns. Observe that there are several times where the heading changes for a short time before reverting back to the previous long-term value. These course corrections have the impact of translating the flight path and then returning to the previous ground velocity vector: in effect they are a kind of S-turn manoeuvre. If one or more of these corrections occurs between measurements then the most likely paths can be biased because there are no measurements to hint that the manoeuvres have occurred and the S-turn trajectory is less probable under the dynamics model than a constant angle path.

Figure 9.6 shows the pdf from the filter: the pdf is multi-modal with three main peaks that are somewhat blurred together. There was a heading change just before the last measurement and the lack of future data makes it impossible to resolve exactly what manoeuvre led to the change in range rate. One of the peaks of the pdf is clearly centred close to the true location.

9.4 9M-MRO 7 March 2014 Beijing to Kuala Lumpur

This flight is the MH371 route that was flown by the accident aircraft 9M-MRO on the morning of 7 March 2014 and is the return flight from Beijing back to Kuala Lumpur. Figure 9.7 summarises the features of the flight: there were three altitude changes and two main heading changes, the first of which was almost immediately after the start of the validation segment. This flight does not contain the S-turn manoeuvres that were present in the previous flight. In addition to the altitude changes the Mach number of the aircraft changed from 0.83 to 0.82. Each of these leads to a change in air speed. This flight contained several anomalous BTO measurements that were corrected using the empirical adjustment described in Chap. 5.

Figure 9.8 shows the pdf output from the filter; the true ACARS aircraft location is again under the main peak of the pdf. The peak is more spread because the altitude changes and Mach change modify the radial speed between the aircraft and the satellite. The resulting BFO measurements can also be explained by course changes: the aircraft could change speed or it could turn slightly. The BFO measurement is not informative enough to discriminate strongly between these and there is not enough subsequent data to see which is more consistent with BTO progression.

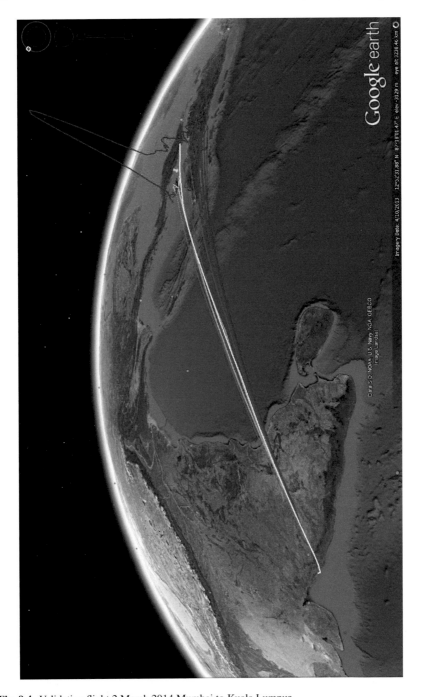

Fig. 9.4 Validation flight 2 March 2014 Mumbai to Kuala Lumpur

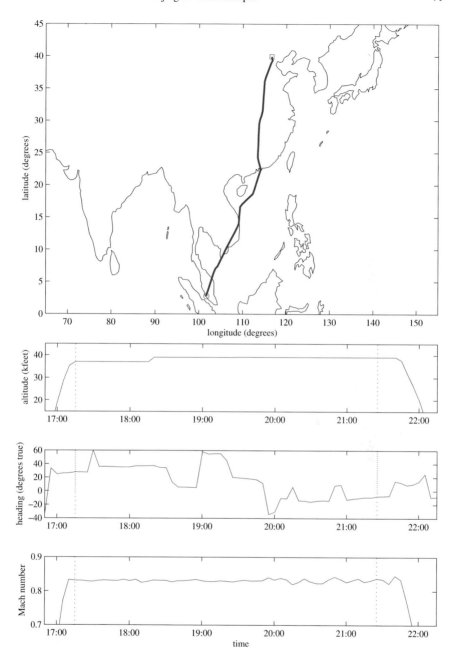

Fig. 9.5 Validation flight 6 March 2014 Kuala Lumpur to Beijing. *Vertical dotted lines* show the start and end times of the segment used for validation

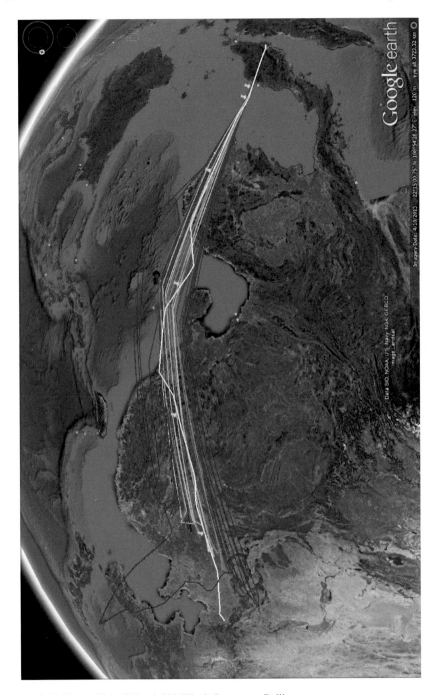

Fig. 9.6 Validation flight 6 March 2014 Kuala Lumpur to Beijing

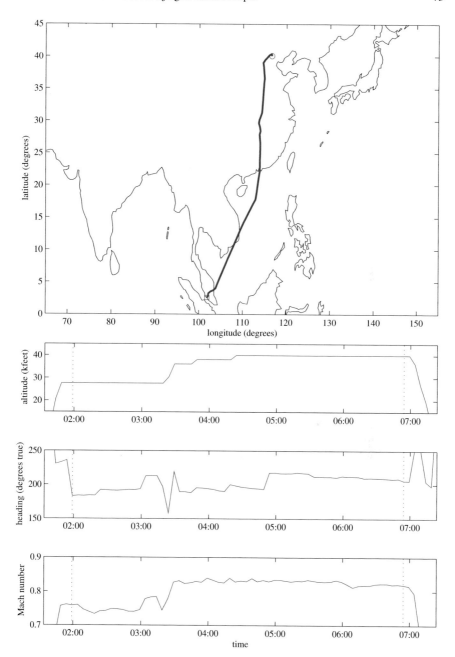

Fig. 9.7 Validation flight 7 March 2014 Beijing to Kuala Lumpur. *Vertical dotted lines* show the start and end times of the segment used for validation

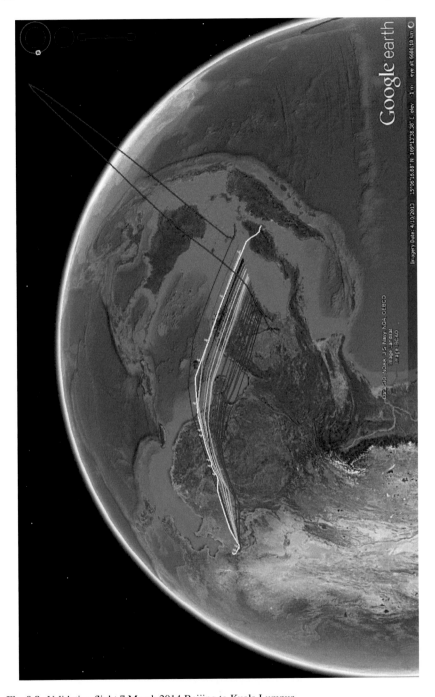

Fig. 9.8   Validation flight 7 March 2014 Beijing to Kuala Lumpur

9.5 7 March 2014 Kuala Lumpur to Amsterdam

This flight was from Kuala Lumpur to Amsterdam and is the same flight path as the first validation flight but with a different aircraft. Figure 9.9 summarises the features of the flight: the aircraft climbs with a sequence of vertical manoeuvres and there is a large S-turn manoeuvre near to the end of the analysed flight segment.

Figure 9.10 shows the pdf output from the filter. The true ACARS aircraft location is under the main peak of the pdf but in this case the true location is lower in the tails than in the other cases. The numerical results that follow in Sect. 9.7 show that this flight had the worst overall performance of the validation flights, although, as discussed in the next section, for each subset of measurements, the final location is within the region containing 85 % of the probability distribution, i.e., the highest posterior density (HPD) interval, discussed further in Sect. 9.7.2.

9.6 7 March 2014 Kuala Lumpur to Frankfurt

The final validation flight was from Kuala Lumpur to Frankfurt. Figure 9.11 summarises the features of the flight. It shows the full flight path, but the communications satellite changes part way through and the test section finishes where the box is marked on the map. No Mach information was available for this flight. There was a large heading deviation mid-flight, but the aircraft eventually reverted back to the earlier heading: this kind of compound manoeuvre is difficult for the filter to characterise. This flight also contained outlier BFO measurements.

Figure 9.12 shows the pdf output from the filter. The performance on this flight is quite similar to the Kuala Lumpur to Amsterdam flights. The filter has again identified ambiguous paths due to the relative geometry.

9.7 Quantitative Analysis

The examples above present a qualitative measure of performance but a more rigorous objective measure is required to provide a statistical assessment of the filter output. So far we have been satisfied that the true location has been in an area of reasonable support for the pdf, but is the spread of the pdf appropriate and is the mode of the distribution biased? Answers to questions such as these require a much larger ensemble of test data. However, it has not been feasible to collect the required test measurements for dozens of different flights. In order to increase our confidence in the performance for the relatively small set of flights that is available, multiple communication measurement sets were generated for each flight by randomly selecting individual R1200 messages from the communication logs of each flight.

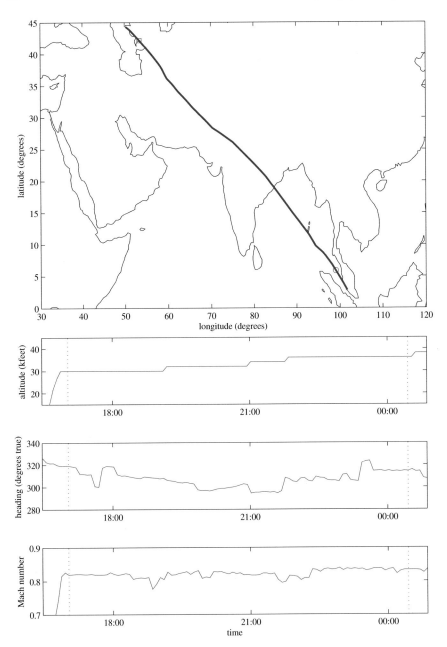

Fig. 9.9 Validation flight 7 March 2014 Kuala Lumpur to Amsterdam. *Vertical dotted lines* show the start and end times of the segment used for validation

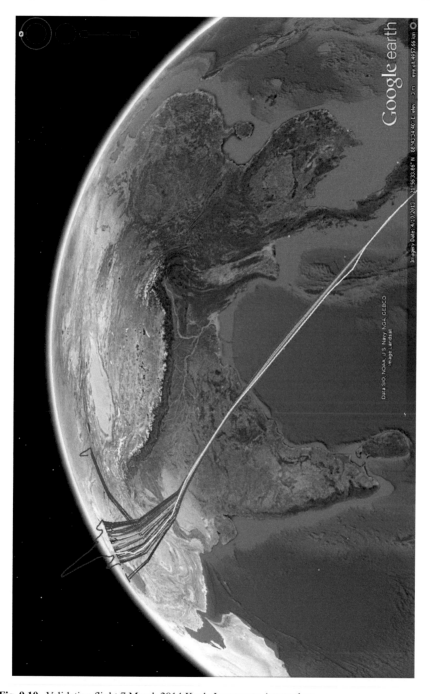

Fig. 9.10 Validation flight 7 March 2014 Kuala Lumpur to Amsterdam

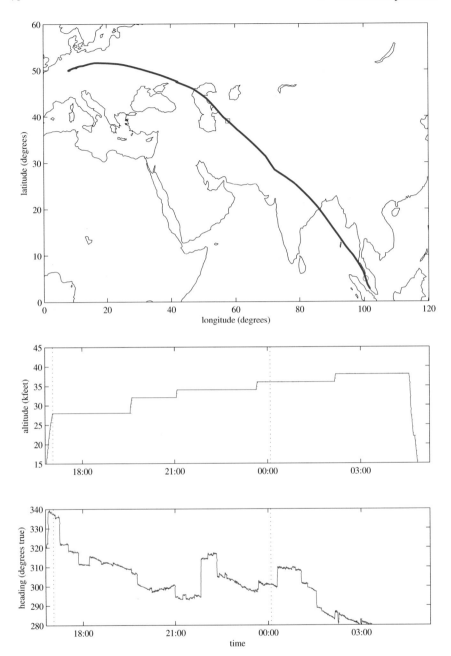

Fig. 9.11 Validation flight 7 March 2014 Kuala Lumpur to Frankfurt. *Vertical dotted lines* show the start and end times of the segment used for validation

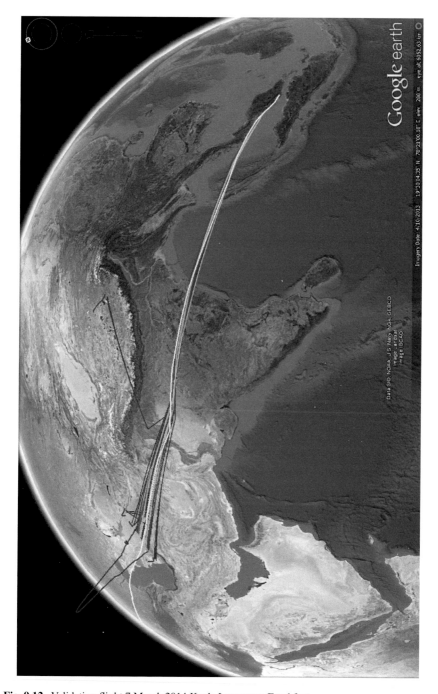

Fig. 9.12 Validation flight 7 March 2014 Kuala Lumpur to Frankfurt

The selection process was repeated 10 times for each flight and these 10 measure-ment sets are treated as independent Monte Carlo random trials for a fixed true air-craft trajectory. As discussed in Chap. 5, the BFO measurement errors are not truly independent over short time periods, which somewhat compromises the assumed independence. However, the common geometry of multiple sets from a single flight is the dominant source of correlation amongst single-flight predictions.

We now briefly review the method used to select individual messages and the performance measure used for this analysis. The chapter concludes with numerical results from these sixty measurement sets.

9.7.1 Measurement Selection

The start and end time for analysis was manually selected for each flight. These times were chosen to exclude ascent from take-off and descent to landing as well as to avoid turns that were very close to either end point. Once these times were determined, the individual measurements were selected using a heuristic randomised process. The intent of this process was to avoid manual selection bias and to create measure-ment sets that emulate the data available for the accident flight. Measurements were selected recursively.

Let t_{k-1} denote the measurement time for the previous measurement; t_0 is the man-ually selected starting time. Each measurement has a collection time labelled t_j, for $j \in \{1, \ldots, J\}$, where J is the total number of measurements in the communication log. The first measurement was selected by assigning a probability

$$p_j(0) = P(0)^{-1} \exp\left\{-\frac{1}{2\sigma^2}\left(t_j - t_0\right)^2\right\}, \tag{9.1}$$

$$P(0) = \sum_{j=1}^{J} \exp\left\{-\frac{1}{2\sigma^2}\left(t_j - t_0\right)^2\right\}, \tag{9.2}$$

where σ was chosen to be 15 min. The selected measurement was then chosen by taking a single multinomial draw on the probability vector $p(0)$. This selection prefers measurements closer to the start time. Subsequent measurements were chosen with a mean time spacing of 1 h. Let $l(i)$ index the measurement chosen as the ith in the sequence. A probability vector for the $(i + 1)$th measurement was defined as

$$p_j(i+1) = \begin{cases} P(i+1)^{-1} \exp\left\{-\frac{1}{2\sigma^2}\left(t_j - t_{l(i)} - 1\right)^2\right\}, & j > l(i) \\ 0, & j \leq l(i), \end{cases} \tag{9.3}$$

$$P(i+1) = \sum_{j=l(i)+1}^{J} \exp\left\{-\frac{1}{2\sigma^2}\left(t_j - t_{l(i)} - 1\right)^2\right\}. \tag{9.4}$$

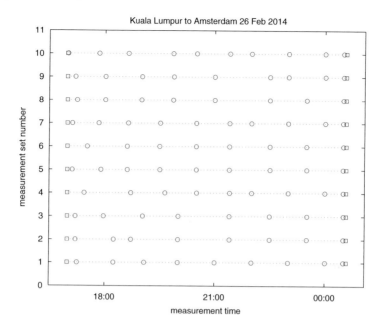

Fig. 9.13 Example measurement timings for a single flight

Measurement $(i + 1)$ is again selected using a single draw on a multinomial distribution defined by $p_j(i + 1)$. The process concludes when the measurement selected occurs after the desired end time: this measurement is discarded.

Figure 9.13 shows an example of the measurement times for the ten different sets generated for the Kuala Lumpur to Amsterdam flight on 26 February 2014. Squares are used to mark the initialisation time and the final time point, neither of which have measurements. The measurement times are marked with circles. Each row is a realisation of the measurement selection process. Some measurements are used by more than one of the sets. The number of measurements selected varies between eight and ten, the duration of the flight segment is approximately seven hours and 35 min: seven one-hour spaces would lead to eight measurements in seven hours.

9.7.2   Performance Measure

In the object tracking literature it is common to use accuracy measures to quantify tracking performance, for example [7]. Accuracy measures quantify how well the estimates from the tracker match the truth. The most frequently used accuracy measure is root-mean-square (RMS) error, which is typically the average geometric distance between the true object position and the tracker estimated position. The requirement for MH370 is a search region, not a point estimate, so RMS is not applicable. The other common accuracy measure is the Normalised Estimation Error Squared (NEES). This is defined as the expectation of the inner product of the estimation error with itself, normalised by the estimator covariance. For a scalar, this is the mean squared error divided by the filter covariance estimate. Whereas RMS quantifies how accurately the filter finds the centre of mass of a distribution, NEES quantifies how accurately the filter estimates the spread of a distribution. NEES inherently assumes a uni-modal distribution. Again, NEES is based on an assumed Gaussian system with a point estimate and covariance estimate. It is not an appropriate measure for the multi-modal pdf produced by the filter in this application. Instead, the statistical performance of the filter output was quantified by measuring the highest posterior density (HPD) interval at the true aircraft location. The HPD interval is defined as the spatial region for which the filter output pdf is at least as high as the value at the true location. Figure 9.14 shows an example of this process for a scalar random variable x with a Gaussian mixture pdf $p(x)$. The two components are equally weighted, one with mean 2 and variance 0.25 and the other with mean 5 and unit variance. Supposing that the truth in this case was $x = 6$, the HPD interval is shaded in red and corresponds to the regions in x for which $p(x) \geq p(6)$. Because the distribution $p(x)$ has two modes and the value of $p(6)$ is between the lower peak and the intermediate minimum, the HPD is composed of two intervals. If the truth had been 2.5 instead then only one region around the higher peak at 2 would be in the HPD and if the truth were 8 then almost all of the pdf would be in the HPD region.

The integral of the pdf over the HPD interval corresponds to the cumulative probability that a random sample from the distribution is more likely than the truth point. If the integral is close to unity, then the HPD interval contains most of the

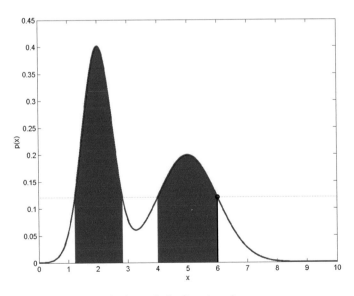

Fig. 9.14 Highest posterior density interval of a Gaussian mixture

support of the pdf, that is the truth point is at a very low part of the pdf. Alternatively, if the integral under the HPD is close to zero then only a small portion of the event space is more likely than the truth point.

Mathematically, the HPD integral is given by

$$h\left(\mathbf{x}^{\text{truth}}; p(\mathbf{x})\right) = \int_{\mathbf{x}: p(\mathbf{x}) \geq p\left(\mathbf{x}^{\text{truth}}\right)} p(\mathbf{x})d\mathbf{x}, \qquad (9.5)$$

where $\mathbf{x}^{\text{truth}}$ is the true aircraft location and $p(\mathbf{x})$ is the filter output pdf. In the discussion that follows, we abbreviate as $h \equiv h\left(\mathbf{x}^{\text{truth}}; p(\mathbf{x})\right)$ the random variable derived by transforming the random variable $\mathbf{x}^{\text{truth}}$ using (9.5).

If the truth values were indeed random samples from the filter output pdfs, then it is relatively easy to show that the distribution of h would be uniform on the interval $[0, 1]$.[1] If integrals tend to be clumped closer to zero then the pdfs being assessed are pessimistic: the tails decay too slowly and the coverage of the pdf is too broad. If the integrals tend to be clumped closer to unity then the truth is always in the tails and the pdfs being assessed are overly optimistic. For the MH370 search definition we prefer a conservative pdf that is a little pessimistic, in order to minimise the chance of excluding the true aircraft location. Provided the search zone defined can be feasibly measured it is better to make this region a little too large and improve the likelihood that the truth is contained.

For each flight we have only ten different measurement sets so it is not feasible to construct a sensible estimate of $p(h)$. Instead we plot an estimate of its cumulative distribution and compare it with the line $y = x$, which is the cumulative distribution of a uniform random variable. If the h values are relatively small then the empirical cumulative distribution function (cdf) will rise more quickly than the reference and the curve will be above it. Conversely if the values are relatively large then the empirical cdf will rise slowly and the curve will be below the reference.

9.7.3 Results

Figure 9.15 shows the empirical cdf derived for each validation flight separately. This shows that the results within a single flight are quite correlated because the filter performance is dependent on geometry. For the Mumbai to Kuala Lumpur, Kuala Lumpur to Beijing and Beijing to Kuala Lumpur flights the h values are generally small but not close to zero. This indicates that the spread of the filter pdf is too high

---

[1]To see this, let $Y = p(\mathbf{x})$, i.e., the random variable obtained by applying the random value $\mathbf{x}$ to its pdf. Then the cumulative distribution function (cdf) of Y, $F_Y(y) = P(Y \leq y)$ is one minus the HPD value in (9.5). It is well-known that, assuming continuity and monotonicity of the cdf, the random variable obtained by passing a random value through its cdf is uniform on the interval $[0, 1]$ (e.g., [13]), and if Y is uniform on $[0, 1]$, then so is $1 - Y$. The necessary assumptions are satisfied if the pdf $p(\mathbf{x})$ contains no non-zero flat regions and no Dirac delta components.

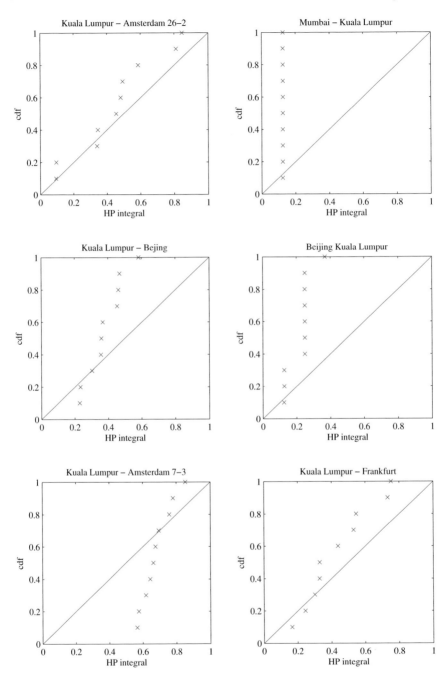

Fig. 9.15 Cumulative density plots for individual validation flights. *Crosses* show individual data-set results and the *solid line* shows the theoretical result for independent samples

and that the peak of the pdf is biased. The bias occurs because the flights make small manoeuvres that are unobservable by the filter. For example, Fig. 9.5 shows that in the Kuala Lumpur to Beijing flight the aircraft made a number of heading changes that lasted for only a short time before the heading reverted back to its previous value. The minor course corrections result in a displacement in position. The filter will sample these paths but their dynamics are less likely than paths without a manoeuvre. For these flights the mode is not a reliable indicator of the true aircraft location but a fairly tight interval is.

In the longer Asia to Europe flights the h values tend to spread between 0.25 and 0.8. Again there is bias due to the repeated geometry and very large values are not observed because the model allows manoeuvres that are more dynamic than what occurred in the actual flights and this spreads the pdf.

Figure 9.16 combines all of the trials into a single h cumulative distribution. In this plot the two different groups of flights are apparent: there is an initial very sharp rise due to the contributions of the intra-Asia flights and then a gradual climb from the Asia to Europe flights.

Overall the results show that for all of the flights and measurement combinations tested the true aircraft location was inside a 85 % confidence region of the pdf. That is, the largest h value observed was approximately 0.85. This means that the pdf estimates are conservative. The spread of the estimated pdf is wider than the spread of true values. This occurs for two reasons: firstly, the aircraft dynamic

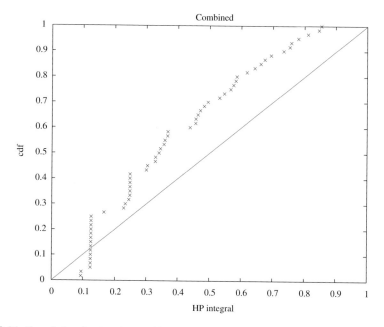

Fig. 9.16 Cumulative density plot combined over all validation flights. *Crosses* show individual data-set results and the *solid line* shows the theoretical result for independent samples

model provides more flexibility than is typically used; for example, in most commercial flights, smaller turns are more likely than turns of 90° or more. Secondly, the assumed measurement variances were deliberately inflated to be pessimistic, as discussed in Sect. 5.3. Given that the accident flight was not a typical commercial flight, the dynamic model should not be exactly matched to typical commercial flights. A somewhat conservative pdf in this case is desirable so long as the pdf does not spread over an area that is unreasonably large to search.

Open Access This chapter is distributed under the terms of the Creative Commons Attribution-NonCommercial 4.0 International License (http://creativecommons.org/licenses/by-nc/4.0/), which permits any noncommercial use, duplication, adaptation, distribution and reproduction in any medium or format, as long as you give appropriate credit to the original author(s) and the source, a link is provided to the Creative Commons license and any changes made are indicated.

The images or other third party material in this chapter are included in the work's Creative Commons license, unless indicated otherwise in the credit line; if such material is not included in the work's Creative Commons license and the respective action is not permitted by statutory regulation, users will need to obtain permission from the license holder to duplicate, adapt or reproduce the material.

Chapter 10
Application to the MH370 Accident

The previous chapters have constructed a Bayesian method for estimating commercial aircraft trajectories using models of the information contained in satellite communications messages and of the aircraft dynamics. This chapter applies the estimator to the accident flight. We show the pdf at the final message time 00:19 and perform several tests that provide confidence in the robustness to the model parameters and measurement characterisation. Probable end of flight dynamics are used to convert the pdf in the air at 00:19 into an ocean search zone. As a further measure of robustness, we show that the model can be initialised at the end of the initial ascent, as was done in the validation flights, and the output pdf remains in the same broad geographic region, even without using the subsequent primary and secondary radar data.

Table 10.1 summarises the available measurement data and the BTO measurement standard deviation, when applicable, since this depends on the message type. The two anomalous R1200 messages and the R600 message occurred during transient phases of operation for the SATCOM equipment so the BFO values reported for these times cannot be used. The BTO measurement variance is also inflated to account for the uncertainty in the correction term.

10.1 The Filter Applied to the Accident Flight

This section summarises the results of applying the variable rate particle filter to the data available for the accident flight. We consider two different treatments of the measurements. Figure 10.1 shows the Google Earth pdf representation for the case where only BTO measurements were used. As with the validation flights, the colour coded paths show the most likely routes to each latitude. Clearly this distribution is multimodal. Based on BTO data, it can be deduced that the aircraft must have manoeuvred in some way between 18:28 and 19:41. However the BTO measurements

© Commonwealth of Australia 2016
S. Davey et al., *Bayesian Methods in the Search for MH370*, SpringerBriefs in Electrical and Computer Engineering, DOI 10.1007/978-981-10-0379-0_10

Table 10.1 SATCOM messages used for MH370 analysis

Time (UTC)	Measurement type	BTO	BTO std. dev. (μs)	BFO
18:25:34	Anomalous R1200	Y	43	N*
18:28:05, 18:28:14	R1200	Y	29	Y
18:39:55	C-channel	N	–	Y
19:41:02, 20:41:04, 21:41:26, 22:41:21	R1200	Y	29	Y
23:15:02	C-channel	N	–	Y
00:10:59	R1200	Y	29	Y
00:19:29	R600	Y	63	N*
00:19:37	Anomalous R1200	Y	43	N*

Measurements marked with an asterisk are available but cannot be used as discussed in the text

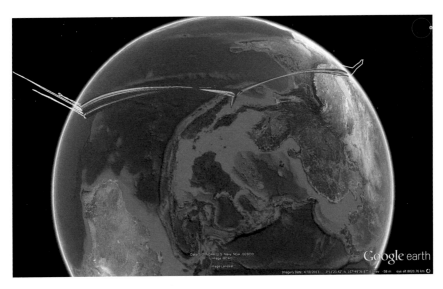

Fig. 10.1 Set of paths from 18:02 to 00:19 using only BTO measurement weighting (i.e., not using any BFO measurements)

are consistent with two options: heading broadly North or broadly South. Many of the paths turn very early, almost immediately after initialisation. This is because the BTO arc at 18:25 is not consistent with the radial speed prescribed by the initial speed and direction. The 18:22 radar point, while potentially inaccurate, implies that no turn had occurred at that stage and that the speed may have changed. Due to the questions surrounding its accuracy, the filter does not use this radar point, so it postulates early direction changes as well as early speed changes to match the 18:25 BTO. A feature of most of the paths is that they do not choose to make subsequent turns even though there is no BFO data to influence angle likelihood. The BTO data itself from 19:41 onwards is consistent with straight and level flight.

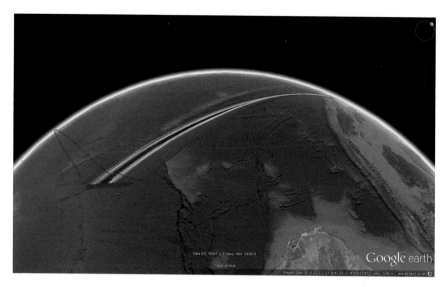

Fig. 10.2 Set of paths from 18:02 to 00:19 using BTO and BFO measurement weighting. *Purple box* shows the November 2015 indicative search region

Figure 10.2 shows how this path set is pruned by including BFO data. The crucial BFO measurements are at 18:28 and 18:39. The first of these is consistent with the same heading as at initialisation, so it implies that the aircraft has not yet turned. In contrast, as discussed in Sect. 5.4, the 18:39 BFO measurement indicates that the aircraft is moving roughly South at this time. Between the two, these measurements restrict the time of the turn to a window between 18:28 and 18:39. This is reflected in the set of surviving paths. Paths that went very far South had to turn earlier and have been rejected as have all of the paths heading to the North towards Asia. The plot also shows the indicative search region as at November 2015, displayed as a purple box.

The main effect of including BFO data is to resolve the ambiguity about the manoeuvre after 18:28. The BFO data does not significantly change the shape of the Southern mode of the pdf. To show this more clearly Fig. 10.3 plots the two pdfs, without and with BFO data, as one dimensional curves parameterised by latitude. The latitude is a nonlinear function of the position around the BTO arc so these pdfs are distorted slightly, but they clearly show the effect of the BFO data: it selects the mode from a multimodal pdf but it does not significantly change the mode shape. Because of this, including only the C-channel BFO measurement has the same effect and changing the assumed BFO noise standard deviation has no significant effect either unless it could be artificially reduced to less than 1 Hz. We conducted several experiments that varied the assumed BFO noise and which BFO values were included. The results are consistent with the curves in Fig. 10.3.

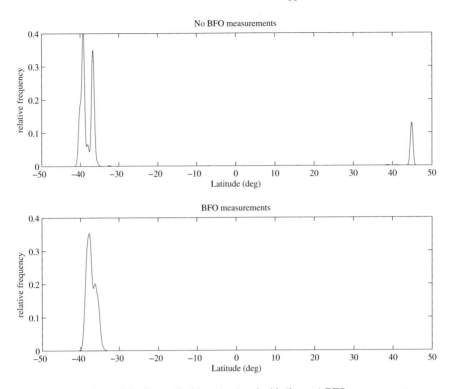

Fig. 10.3 Comparison of the filter pdf without (*top*) and with (*bottom*) BFO measurements

10.2   Manoeuvre Statistics

An interesting question is whether the posterior pdf output from the filter provides any strong estimate of the number and type of manoeuvres after the loss of primary radar coverage. To address this we compare the probability mass of the number of turns for the case of no measurements, i.e., the prior probability from predictions used to create the enormous disc pdf in Fig. 7.4, and the posterior probability after applying the measurements. These are shown in Figs. 7.3 and 10.4 respectively. The posterior distribution does not count turns made between 00:11 and 00:19 because the BFO data at 00:19 is not used (as it is believed to be unreliable); this permits the filter to make superfluous manoeuvres. The histograms were created by counting the number of manoeuvres performed by each particle and then scaling their contribution to the histogram by the particle weight. In the prior probability there is a significant probability of making a very large number of manoeuvres. The particles that contribute to this part of the histogram are those that have selected very short mean manoeuvre times τ.

The histograms show that the posterior number of turns has reduced dramatically, as has the number of speed changes. The likelihood of selecting a sequence of random

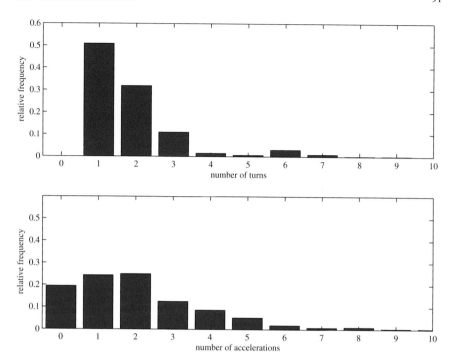

Fig. 10.4 Histogram of posterior number of turns and speed changes

speeds that average out to a suitable overall speed is relatively high. Since there is usually at least an hour between BTO arcs, the filter has time to make several speed changes. Only the average of these speeds influences the position and hence the BTO. As discussed several times, the BFO is sufficiently uninformative that the instantaneous velocity at the measurement time is only loosely constrained. However, it is unlikely that a sequence of random turns will "cancel each other out," and the filter finds very few paths that string together multiple turns when a straight path would suffice. The low support for turns leads to selection of longer mean manoeuvre times, which in turn suppresses superfluous speed changes.

The turn histogram indicates that around half of the paths made more than a single turn. This would appear to be of interest but is in fact misleading. Figure 10.5 shows the posterior distribution of angle as a function of time. The greyscale image shows the probability density at each time slice; the darkest points are most likely. The solid line shows the mean and one-sigma error bars indicate the standard deviation. Note that the horizontal scale is measurement index, not time, but time values are used to label the measurements to give perspective. The angle plot shows that there are a number of particles that make a very small turn before 18:28 and then follow this with a second turn between 18:28 and 18:39. The result is very similar to making a single turn through the same total angle. Similarly, there are a number of particles that make a turn before 18:39 and then a second very small turn immediately afterwards. Again, this is approximately equivalent to a single turn through the same total angle.

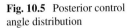

Fig. 10.5 Posterior control angle distribution

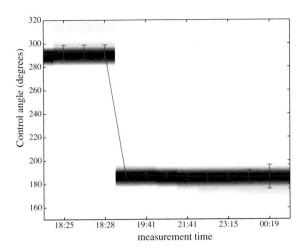

The apparent large number of double turns is really just a single turn being broken into two segments. There appears to be very few genuine turns later in the flight.

Figure 10.6 shows the marginal distribution of the Mach number between 18:02 and 00:19. The prior is uniform in Mach between 0.73 and 0.84 but the posterior strongly prefers the higher speed part of this range. This is approximately consistent with the most likely speeds for the aircraft over long durations.

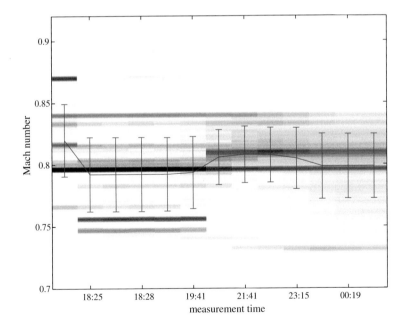

Fig. 10.6 Posterior speed distribution

10.3 Residual Measurement Errors

Another measure of the quality of an estimate is the statistical behaviour of the
measurement residual errors. If the estimates were exactly the truth then the residual
measurement errors would be the actual measurement noise. So these residuals should
be uncorrelated, zero mean, and spread consistently with the measurement noise
variance. Figure 10.7 shows the marginal distributions of the residual measurement
errors. The filter predicts the BTO and BFO at the measurement times based on
the estimated position and velocity and the residual is the difference between these

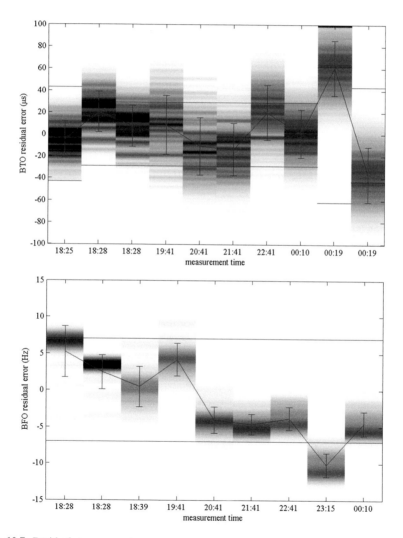

Fig. 10.7 Residual measurement error

predictions and the measured values. Each sample is weighted according to its overall likelihood and the combination gives the marginals in Fig. 10.7. The horizontal axis is again the measurement index and the vertical axis is residual error, dark points have high probability, and the error bars show the one-sigma limits. The figure also overlays the one-sigma lines for the assumed measurement noise. For BTO this varies with the message type, for BFO it is a fixed ± 7 Hz. There is no BTO measurement for the C channel communications that occurred at 18:39 and 23:14. The BFO has been discarded at 18:25 and 00:19 since these readings are thought to be unreliable. The plots only show residuals when a valid measurement is available.

For the BTO residual there are two larger residuals on the final two measurements. These measurements are actually very close together in time: the first is an R600 message at 00:19:29 and the second is an R1200 message at 00:19:37. The reason why both of these measurements show large residuals is that they are not consistent with each other: the residuals have opposite signs, reflecting that one measurement is longer in range than the prediction and the other is shorter. In the absence of a reason to prefer one over the other, we use both measurements and let the filter find paths that are the best statistical fit. The BFO residuals are statistically consistent with the empirical error model.

10.4   Posterior Distribution of Manoeuvre Time Constant

The inference procedure that has been developed samples the manoeuvre time constant, which is a static parameter. In the case of conventional particle resampling, this is not advisable, as subsequent resampling steps generally leave few distinct values of the static parameter. The recursive resampling process described in Chap. 8 is less prone to this difficulty, as many independent sample paths are retained.

The sample support of the posterior distribution of the manoeuvre time constant is demonstrated in Fig. 10.8. The diagram clearly shows that the distribution is non-degenerate. The figure compares the posterior distribution from the filter, which is shown as bars, with the Jeffreys prior, which is shown as a red line. The Jeffreys prior prefers lower mean manoeuvre times but the data does not support these. The posterior is dominated by longer mean manoeuvre times, with 97 % of the distribution having a mean time between manoeuvres of more than one hour, and 83 % having a mean time more than two hours.

10.5   End of Flight

The output of the particle filter is an estimate of the pdf of the aircraft state at 00:19. The aircraft was still in the air at this time and a model is required to describe the distribution of how it may have descended. This has been primarily the responsibility of the ATSB and the other members of the search team. A discussion of the different

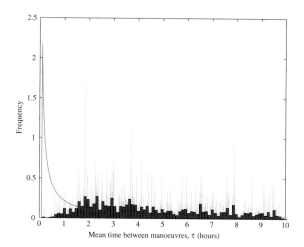

Fig. 10.8 Marginal posterior distribution of manoeuvre time constant, τ. *Red line* shows Jeffrey's prior, as described in Sect. 7.2.1. *Bars* show histogram of samples weighted by measurement likelihood, on bins of width 0.12 h. *Gray line* shows histogram on finer bins of width 0.012 h

methods used to model the potential motion is presented in [5]. The model for aircraft motion after 00:19 leads to a prioritisation of the search along and around the final BTO arc.

Quantitatively, the model of descent defines a transition distribution $p(\mathbf{x}_{\text{final}}|\mathbf{x}_K)$, which describes the probability density of the final state given the state at time 00:19. In effect this acts as a kernel to spread the distribution at 00:19 via the expression

$$p\left(\mathbf{x}_{\text{final}}|\mathbf{Z}_K\right) = \int p(\mathbf{x}_{\text{final}}|\mathbf{x}_K) \sum_{p=1}^{P} w_K^p \delta(\mathbf{x}_K - \mathbf{x}_K^p) d\mathbf{x}_K = \sum_{p=1}^{P} w_K^p\, p(\mathbf{x}_{\text{final}}|\mathbf{x}_K^p)$$

$$(10.1)$$

The analysis in [5] leads to a probable scenario where the aircraft ran out of fuel at some time between 00:11 and 00:19. The final satellite communications message could be due to the modem rebooting under auxiliary power. Under this hypothesis, the aircraft engines were already unpowered at 00:19. The spread of the kernel function is then determined by the distance over which the aircraft could have moved, which depends on whether or not the aircraft was under human control during this period.

Flight simulator studies of uncontrolled descents have shown a high likelihood of the aircraft reaching zero altitude within 15 nm of the beginning of descent [5]. However, the beginning of descent is not known. It is possible for the aircraft to have travelled farther, especially if a human was controlling the aircraft. As an indicative kernel, and following advice from the ATSB, a uniform disc of radius 15 nm with a Gaussian drop off with standard deviation 30 nm beyond this was chosen; this represents the accident investigators' assessment of the likely scenarios. Figure 10.9 shows a radial slice through this kernel function. The kernel was convolved with the particle locations at 00:19 to generate the heat map shown in Fig. 10.10. The black region overlaid on the heat-map shows an indicative bounding box on the geographic

Fig. 10.9 Descent kernel

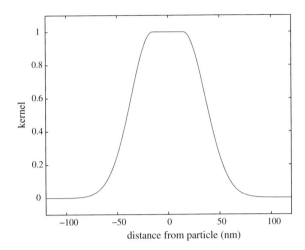

search region. The figure also shows the 99, 95 and 90 % confidence regions, i.e., the smallest geographic area containing each percentage of the pdf.

10.6 Earlier Initialisation

As a further level of validation on robustness, a separate run of the filter was performed initialising using the final ACARS report. This run is not useful for defining a search region because it ignores important data from the primary and secondary radar. However, the result does illustrate the ability of the method to address flight paths with several large turns. Figure 10.11 shows the pdf derived from the 17:01 initialisation. The pdf is overlaid with the one that is presented earlier in this chapter that uses all of the available information. Two versions of the 17:01 initialisation are presented. The first, which ignores the primary radar data, is shown in red in the upper diagram. The second, shown in the lower diagram, re-weights the paths by treating the 18:02 radar point as a measurement with 1° standard deviation in latitude and longitude. This weighting is rather loose but the samples are limited and none of them are sufficiently close to the radar point to permit a tighter match. The filter has given a similar output even though the initialisation point was heading North and there are two very significant manoeuvres reported by the radar in the first hour. This provides further confidence in the ability of the filter to characterise aircraft motion even in complicated situations. The sequence of turns performed by MH370 between 17:01 and 18:02 is much more dynamic than any experienced in the validation flights. When the radar data was ignored the filter found a cluster of paths that travel a significant distance to the North before turning prior to 18:39. These paths conclude at the Northern edge of the search region because they travel too far to the North early on, in a manner which is inconsistent with the primary radar data. In contrast, the

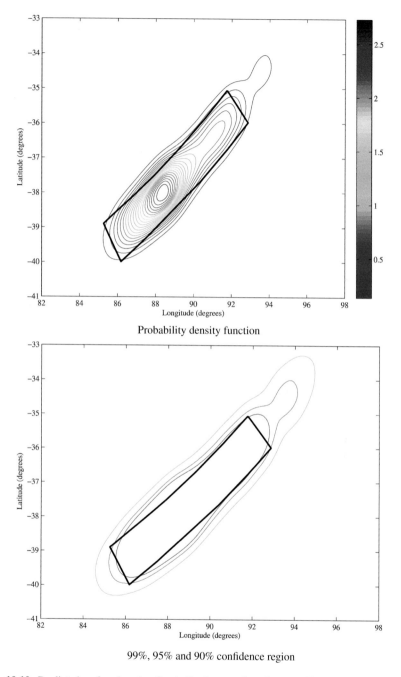

Probability density function

99%, 95% and 90% confidence region

Fig. 10.10 Predicted surface location for: indicative search region as at November 2015 marked with *solid line*

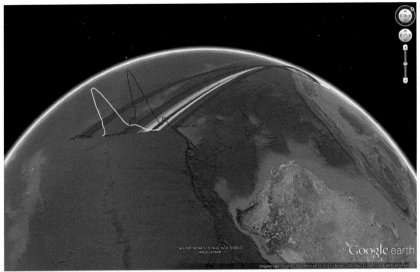

17:01 initialisation with no radar data (red) compared with the 18:02 radar initialisation (white)

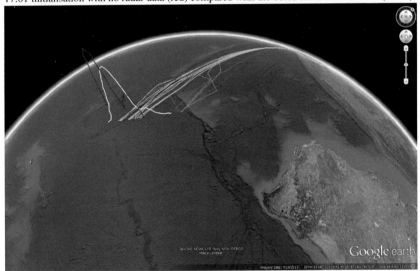

17:01 initialisation weighted by proximity to 18:02 radar (red) compared with the 18:02 radar based
initialisation (white)

Fig. 10.11 Probability density from early initialisation

paths that are preferred after re-weighting using the primary radar data travel South
of the radar point and arrive at the Southern edge of the search region. One would
expect that, given further computation, samples would be found between these two
that would match the 18:02 radar point better and agree better with the search region.

10.7 Cost Index

Another autopilot mode for control of air speed is Cost Index, as discussed in Sect. 6.2.1. To recap, under this mode the autopilot dynamically selects a speed in order to optimise a criterion which trades fuel consumption and travel time according to the Cost Index value entered by the pilot [34]. An additional experiment was performed in which the aircraft was permitted to enter into the Cost Index mode at a random time, remaining in that mode from that time forward. The Cost Index value was randomly sampled between 0 and 100, and lookup tables provided by Boeing were used to determine a speed based on the sampled value of altitude, and an estimate of the aircraft weight, which varies over time as fuel is expended.

  The result of this experiment is shown in Fig. 10.12. The white curve shows the pdf calculated through the model using Mach number, as shown in Fig. 10.2, while the red curve shows the pdf calculated using the Cost Index model. The figure shows that the use of Cost Index does not significantly change the result; the distribution is slightly more compact than in the case of the constant Mach number model. Since either mode is possible, the broader result in Fig. 10.2 is preferred for determining the search region.

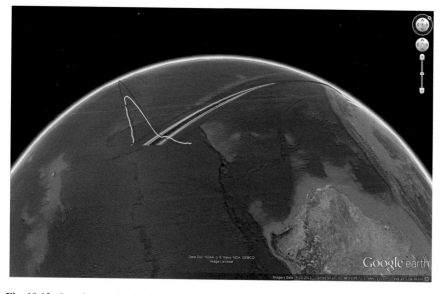

Fig. 10.12 Set of paths obtained using Cost Index mode for controlling air speed (*red*), compared to result using the Mach number model (*white*) as shown in Fig. 10.2

10.8   Other Variations

The results presented in this chapter are a summary of the analysis we have undertaken on the accident flight data. They are far from an exhaustive set of the variations that have been examined. Many other changes to the filter parameters and measurement model were explored but none made a significant impact on the search region. Example changes to the filter include: modifying the Jeffreys prior; using fixed average manoeuvre times; changing the incremental time step; resampling the Jefferys prior to emulate a change in aircraft behaviour; moving the initialisation point; modifying the initialisation variances; forcing a prescribed angle after 18:39; and quantising heading to integer degrees. Example changes to the measurement model include: vetoing selected measurements; using only C-channel BFO measurements; increasing and decreasing the BFO assumed noise variance; replacing the BFO with a "true" BFO using known location for the validation flights; including or excluding particular BTO measurements and R600 messages; and cross validation where each measurement is removed in turn to investigate output robustness.

In our earliest attempts at producing a pdf a much simpler aircraft model was used. For example, under the assumption of a single turn, the parameter vector is relatively short and more conventional estimation methods can be used. However, such a model cannot be applied to any of the validation flights and runs the risk of constraining the estimated path with additional assumptions. The fact that the very general model here chooses to make few turns provides information that would not be available from a single turn geometric model.

The factors that do make a significant difference to the output pdf are the assumed spread of Mach number and the end of flight model. The assumed Mach number range covers the speeds feasible to achieve the required flight endurance time. The lower end of this speed range results in the Northern part of the pdf and the higher end of the speed range results in the Southern part. Restricting the speed to only Mach numbers above 0.8, for example, would contract the pdf to the South. The consequence is that using a smaller speed range within the bounds already modelled leads to a subset of the search zone. If a different end of flight model is assumed the general consequence is to spread the search zone over a larger area. Simulations have predicted that the maximum distance that the aircraft could have glided under human control is approximately 100 nm after 00:19 [5]. The search zone that this scenario would imply is very much larger.

Open Access   This chapter is distributed under the terms of the Creative Commons Attribution-NonCommercial 4.0 International License (http://creativecommons.org/licenses/by-nc/4.0/), which permits any noncommercial use, duplication, adaptation, distribution and reproduction in any medium or format, as long as you give appropriate credit to the original author(s) and the source, a link is provided to the Creative Commons license and any changes made are indicated.

The images or other third party material in this chapter are included in the work's Creative Commons license, unless indicated otherwise in the credit line; if such material is not included in the work's Creative Commons license and the respective action is not permitted by statutory regulation, users will need to obtain permission from the license holder to duplicate, adapt or reproduce the material.

Chapter 11
Ongoing Refinement

The previous chapters of this book have built up models for the limited available measurement data and for aircraft motion, and then used these models to produce a pdf of the final aircraft latitude and longitude. In principle, one could use this pdf to direct a search and then the very act of searching would provide further measurement data. This chapter describes how the Bayesian method can be used to adapt the position pdf to account for data collected after the accident flight. Two data sources are discussed: the sonar imagery data collected in the search and the discovery of the flaperon on Reunion Island.

11.1 Updating the Distribution Using Search Results

The measurements collected as part of the search can be treated in the same mathematical framework as the communications data. This was used, for example, in the search for AF447 [40]. In this case the aircraft is no longer moving so the prediction stage becomes degenerate and the predicted pdf is the same as the previous posterior pdf. The result of the search can be summarised by the probability that the cumulative search effort would have detected the aircraft at any particular location, $P_D(\mathbf{x})$. The posterior pdf given the search effort (denoted S) then becomes:

$$p(\mathbf{x}_{\text{final}}|S, \mathbf{Z}_K) \propto [1 - P_D(\mathbf{x}_{\text{final}})]p(\mathbf{x}_{\text{final}}|\mathbf{Z}_K), \qquad (11.1)$$

where the constant of proportionality is determined to ensure that the posterior integrates to unity. If a particular area $\mathcal{A}$ is searched with a constant probability of detection P_D, the probability of finding the aircraft is

$$P(\text{find during search of } \mathcal{A}) = P_D \int_{\mathcal{A}} p(\mathbf{x}_{\text{final}}|\mathbf{Z}_K) \, d\mathbf{x}_{\text{final}}. \qquad (11.2)$$

© Commonwealth of Australia 2016
S. Davey et al., *Bayesian Methods in the Search for MH370*, SpringerBriefs
in Electrical and Computer Engineering, DOI 10.1007/978-981-10-0379-0_11

In the analysis used for prioritising the search for flight AF447, the probability of detection for areas searched using side-scan sonar was modelled as 0.9 [40]. This type of analysis could be used, for example, to prioritise whether to revisit a more likely area or to search a less likely area that has not yet been searched.

Based on the quality assurance process which has been implemented in the MH370 search, which includes revisiting items that have been assessed as potential debris, it is considered highly unlikely that the search would fail to detect the aircraft if the correct location is searched. Due to sensor drop-out and terrain masking, there will inevitably be small pockets which are not covered in a first pass of the search, and (11.1) can be used to determine the priority of returning to ensure that these are examined.

11.2   Reunion Island Debris

The apparent lack of debris from the aircraft was a mystery which was resolved in part when a flaperon was discovered washed up on Reunion Island on 29 July 2015, 508 days after the accident, and later confirmed to be from MH370. Two questions arise from this find:

1. What information does the discovery of the flaperon on Reunion Island provide about the final location of the aircraft?
2. What information does the lack of any other debris to date provide about the final location of the aircraft?

Debris information was used to inform the search for AF447 [40], and was proposed for MH370 in the white paper [19]. The most directly relevant source of information on the likely drift of the flaperon is from the Global Drifter Program [32]. The program provides 30 years of data on buoys referred to as drifters, which are regularly deployed in oceans worldwide. The primary goal of the program is to measure ocean currents rather than surface effects. To ensure that the drifter motion is dominated by the ocean current, the drifters are deployed with drogues, which are sea anchors that sit around fifteen metres beneath the ocean surface. If the drogue is detached then there is an observable change in the motion of the drifter [28]. A drifter without a drogue is referred to as an undrogued drifter and is more buoyant than the drifters with an attached drogue. The buoyancy of the flaperon has not been characterised at this time but it is expected that the undrogued drifters better describe its likely motion because the drogues by design cause drifters to move according to deeper currents.

In this section, we derive an updated pdf of the final location of the aircraft based on an analysis of data from the Global Drifter Program. It should be noted that the characteristics of the drifters are not accurately matched to the flaperon, thus there is some uncertainty in the applicability of the results, but it is believed to be the most relevant data source available. The authors gratefully acknowledge Dr. David Griffin (CSIRO, Australia) for providing specialist analysis and advice in support of this effort, as well as the results illustrated in Fig. 11.1. Further information may be found at [17].

The first question is answered by modifying the pdf provided in the previous chapter. This is achieved by developing a likelihood function of the source location of the flaperon using the drifter data. The likelihood is used to calculate the posterior pdf of the final aircraft location given both the information provided by the Inmarsat satellite data and the discovery of the debris.

The second question relates to the potential information provided by the lack of other debris. We consider a statistical framework for quantifying this information but the framework requires parameters that cannot be reliably determined. For this reason we limit ourselves to qualitative conclusions about the lack of other debris.

11.2.1 Update of Final Location Distribution

We start with the form which results from the particle filter: (sampling to incorporate the effect of the descent kernel of Sect. 10.5)

$$p\left(\mathbf{x}_{\text{final}}|\mathbf{Z}_K\right) \approx \sum_{p=1}^{P} w^p \delta\left(\mathbf{x}_{\text{final}} - \mathbf{x}^p\right), \tag{11.3}$$

For a single item of debris, we seek a model which permits us to update (11.3) using the knowledge that the item arrived at a given location and time. For this, we require a transition distribution for how the debris would move over 508 days. Denoting the location after 508 days as $\mathbf{y}$, this transition distribution is $p(\mathbf{y}|\mathbf{x}_{\text{final}})$. Thus, assuming a single item of debris, the kinematic distribution updated by the knowledge of the debris find at location $\mathbf{y} = y$ is:

$$p(\mathbf{x}_{\text{final}}|\mathbf{Z}_K, \mathbf{y} = y) \propto p(\mathbf{y} = y|\mathbf{x}_{\text{final}})p(\mathbf{x}_{\text{final}}|\mathbf{Z}_K)$$

$$\approx \sum_{p=1}^{P} w^p p\left(\mathbf{y} = y|\mathbf{x}^p\right) \delta\left(\mathbf{x}_{\text{final}} - \mathbf{x}^p\right), \tag{11.4}$$

Equation (11.4) can be interpreted as re-weighting each particle $\mathbf{x}^p$ by the likelihood that an item of debris at that location would end up at location $\mathbf{y} = y$.

A field of debris can be modelled as a Poisson point process (e.g., [42]). Suppose that the expected number of debris items is λ, that an item that is washed up on the shore is identified with probability P_I, and that no items are found while they remain in the ocean. Then the likelihood of the first item of debris being found at $\mathbf{y} = y$ given that the end-of-flight location was $\mathbf{x}_{\text{final}}$ is:

$$l(\text{debris}|\mathbf{x}_{\text{final}}) = \exp\left\{-P_I\lambda \int_C p(\mathbf{y}|\mathbf{x}_{\text{final}})d\mathbf{y}\right\}$$

$$\times \exp\left\{-\lambda p(\mathbf{y} = y|\mathbf{x}_{\text{final}})\right\} \lambda p(\mathbf{y} = y|\mathbf{x}_{\text{final}}) \tag{11.5}$$

where the integral is over the coastal region $\mathcal{C}$, i.e., the region where debris would be likely to have been found. This expression could be used to update the model as:

$$p(\mathbf{x}_{\text{final}}|\mathbf{Z}_K, \text{debris}) \propto l(\text{debris}|\mathbf{x}_{\text{final}})\,p(\mathbf{x}_{\text{final}}|\mathbf{Z}_K)$$

$$\approx \sum_{p=1}^{P} w^p l\left(\text{debris}|\mathbf{x}^p\right) \delta\left(\mathbf{x}_{\text{final}} - \mathbf{x}^p\right), \qquad (11.6)$$

Similar to the case of a single debris item, this model re-weights each particle $\mathbf{x}^p$ with the likelihood that a debris field starting from that location would not yield any identified coastal debris within the time other than a single item at $\mathbf{y} = y$. The model could be easily refined to incorporate thinning, i.e., the likelihood that over time, items of debris might sink, as well as the temporal aspect that the debris did not arrive until some 508 days later, by splitting the transition distribution $p(\mathbf{y}|\mathbf{x})$ into a number of intermediate, smaller duration transition steps. However, the required model parameters, are very difficult to quantify with any degree of confidence. Specifically, the expected number of debris items, and the probability that an item washed up would be identified, are assumed to be known. Since these parameters cannot be adequately specified, we rely on a qualitative approach for incorporating information provided by the lack of debris, and use the quantitative approach of (11.4) for incorporating the information on the debris item that was found.

11.2.2 Data from Global Drifter Program

The data available for undrogued drifters are summarised in Fig. 11.1. Each diagram shows drifter trajectories which pass through a region of interest in months between February and April (i.e., the time of year of the accident). The middle figure centres the region of interest on the search zone, while the top figure examines the region to the North, and the bottom figure shows the region to the South. Coloured dots are drawn along each drifter trajectory every 100 days to illustrate time along the trajectory. Trajectory segments coloured in blue indicate water colder than $18\,^\circ\text{C}$, the temperature above which barnacle nauplii settlement and growth is accelerated; evidence of accelerated growth was present on the flaperon [17].

The following observations can be made from the figures:

1. In the top diagram (the region to the North of the search zone), many trajectories head West at quite a fast rate, reaching Madagascar and Tanzania in 300 days or less. The absence of debris being identified in the Western Indian Ocean many months earlier than July 2015 would tend to indicate this region as being less likely.
2. A significant proportion of the trajectories in the middle diagram (the search zone) arrive in the general vicinity of Reunion Island at around 500 days.

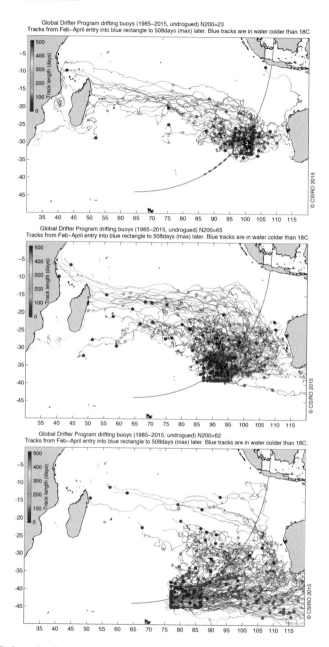

Fig. 11.1 Trajectories from global drifter program. *Top* diagram shows trajectories which pass through the box with latitudes from 27 °S to 33 °S and longitudes from 96 °E to 106 °E. *Middle* diagram shows trajectories which pass through the box from 33 °S to 39 °S and longitudes from 87 °E to 96 °E. *Bottom* diagram shows trajectories which pass through the box from 39 °S to 45 °S and longitudes from 77 °E to 87 °E. *Coloured dots* show time, marking every 100 days. Blue trajectories are in water colder than 18 °C, the temperature above which barnacle nauplii settlement and growth is accelerated. Figure courtesy of Dr. David Griffin, CSIRO; see also [17]

3. The majority of trajectories in the bottom diagram (the region to the South of the search zone) head to the East, towards Australia and New Zealand. Very few head toward the vicinity of Reunion Island.

Thus the discovery of the single flaperon on Reunion Island after 508 days would seem to be more consistent with the existing search region, and less consistent with regions significantly to the North or South.

11.2.3 Posterior Distribution Using Debris Data

Figure 11.1 represents the total undrogued drifter data available passing through each box at the relevant time of the year over the 30 year history of the Global Drifter Program. It is insufficient to construct a transition distribution, hence additional processing was performed to enhance the data set. Specifically, pairs of trajectories that passed close together (possibly in different years) were "joined," i.e., used to create additional synthetic trajectories comprised of the head of one with the tail of the other. A synthetic data set was generated by joining four trajectory segments. In each joining process, each previous trajectory creates around 30 new trajectories.

The joined trajectories provide a sampled representation of the distribution:

$$p_{\text{drifter}}(\mathbf{x}, \mathbf{y}) \approx \frac{1}{P} \sum_{p=1}^{P} \delta(\mathbf{x} - \mathbf{x}^p) \delta(\mathbf{y} - \mathbf{y}^p) \tag{11.7}$$

where $\mathbf{x}$ is the location of the drifter on 8 March 2014, and $\mathbf{y}$ is the location on 29 July 2015. As in (11.4), the distribution of interest is $p(\mathbf{y}|\mathbf{x}_{\text{final}})$ evaluated for $\mathbf{y}$ at Reunion Island, or, pragmatically, integrating over a small region $\mathcal{R}$ surrounding Reunion Island:

$$l(\mathbf{x}_{\text{final}}) = \int_{\mathcal{R}} p(\mathbf{y}|\mathbf{x}_{\text{final}}) d\mathbf{y} \tag{11.8}$$

The conditional distribution $p(\mathbf{y}|\mathbf{x})$ requires normalisation by the prior distribution $p_{\text{drifter}}(\mathbf{x})$; this represents the overall mean density of drifters in the ocean, which is neither spatially uniform nor time invariant. Applying a small kernel $K(\mathbf{x})$ around each sample, the likelihood is evaluated as:

$$l(\mathbf{x}_{\text{final}}) = \frac{\int_{\mathcal{R}} p_{\text{drifter}}(\mathbf{x}_{\text{final}}, \mathbf{y}) d\mathbf{y}}{p_{\text{drifter}}(\mathbf{x}_{\text{final}})} \approx \frac{\frac{1}{P} \sum_{p=1}^{P} K(\mathbf{x}_{\text{final}} - \mathbf{x}^p) + \varepsilon}{\frac{1}{\tilde{P}} \sum_{p=1}^{\tilde{P}} K(\mathbf{x}_{\text{final}} - \tilde{\mathbf{x}}^p) + \varepsilon} \tag{11.9}$$

where $\{\mathbf{x}^1, \ldots, \mathbf{x}^P\}$ is the set of locations for trajectories which pass though the region $\mathcal{R}$ in the right time window, including those synthetically augmented using the aforementioned joining process. $\{\mathbf{x}^1, \ldots, \tilde{\mathbf{x}}^{\tilde{P}}\}$ is the overall set of drifter locations for the time interval from February to April. The ε terms are added for regularisation,

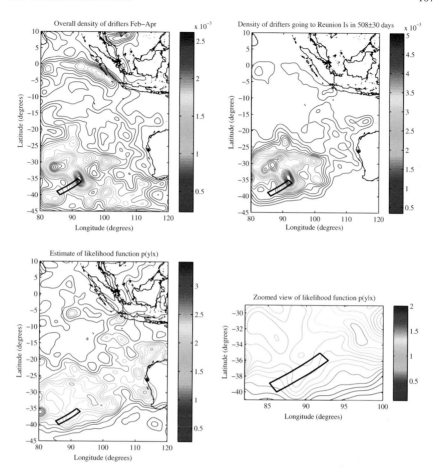

Fig. 11.2 Distributions obtained using undrogued drifter data from Global Drifter Program. *Top-left* diagram shows the overall density of drifters in the months of interest (over all years). *Top-right* shows distribution of synthetically augmented trajectories which pass though the Reunion Island region after 508 days. *Lower-left* diagram shows the quotient of these two distributions, which is the likelihood $p(\mathbf{y}|\mathbf{x}_{\text{final}})$. *Lower-right* diagram shows this same likelihood, zoomed in to the region of interest for the posterior distribution. An indicative search area is marked as a *black rectangle*. Data courtesy of Dr. David Griffin, CSIRO; see also [17]

i.e., preventing very large values in the conditional resulting from the finite sample support of the denominator (we set $\varepsilon = 10^{-4}$).

The result of this analysis is shown in Fig. 11.2, using a fixed kernel of standard deviation $1°$ in latitude and longitude, and evaluating the density estimates using the KDE toolbox [22]. The top left diagram shows the overall spatial distribution of drifters; the indicative search area is again shown in black. The overall distribution clearly has significant spatial variation, particularly in the search area. The density of drifters that pass though the Reunion Island region after the right duration is shown

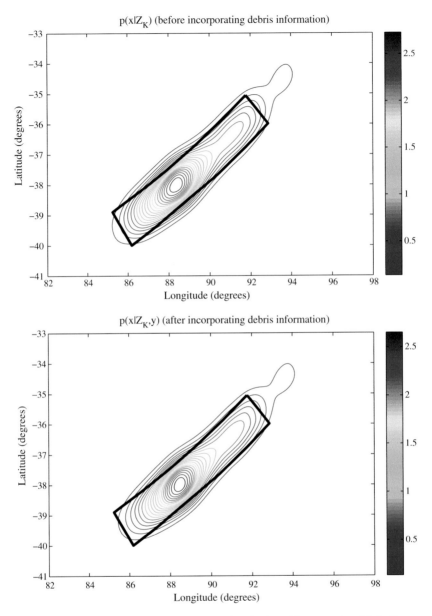

Fig. 11.3 Posterior distribution based on Inmarsat data (*top*), and incorporating debris discovery at Reunion Island (*bottom*)

in the top right. It exhibits spatial variations very similar to the overall variations in the top left. The likelihood function $p(\mathbf{y}|\mathbf{x}_{\text{final}})$ is the ratio of these two densities and is shown in the lower left. Note that large values can occur in areas of limited sample support due to noise. The variation in the distribution of drifters that arrived at Reunion Island is almost completely explained by the variation in the overall starting distribution, therefore the likelihood shows much less spatial variation. The lower right diagram shows a zoomed in region around the search area.

Figure 11.3 shows the posterior distribution based on the Inmarsat data from Fig. 10.10, alongside the updated distribution, which incorporates the information from the debris item located at Reunion Island. The updated distribution is shifted very slightly to the North, but the effect is negligible. This result is not unexpected given the long duration between the accident and the debris discovery.

Open Access This chapter is distributed under the terms of the Creative Commons Attribution-NonCommercial 4.0 International License (http://creativecommons.org/licenses/by-nc/4.0/), which permits any noncommercial use, duplication, adaptation, distribution and reproduction in any medium or format, as long as you give appropriate credit to the original author(s) and the source, a link is provided to the Creative Commons license and any changes made are indicated.

The images or other third party material in this chapter are included in the work's Creative Commons license, unless indicated otherwise in the credit line; if such material is not included in the work's Creative Commons license and the respective action is not permitted by statutory regulation, users will need to obtain permission from the license holder to duplicate, adapt or reproduce the material.

Chapter 12
Conclusions

In this book we have described our Bayesian approach to defining the MH370 search zone. The three ingredients required for the Bayesian approach are

- a prior (defined by the Malaysian military radar)
- a likelihood function describing the relationship between BTO and BFO measurements and the aircraft state vector and
- a model of the aircraft dynamics.

All three are described in detail. To increase confidence in the models and process developed, validation has been performed using previous flights of the accident aircraft, and data from other flights in the air at the same time as the accident flight. In all cases the true location aligns with that predicted by the Bayesian analysis. All validation trajectories have significant numbers of speed, heading and altitude changes, which are successfully captured by the model. This is in stark contrast with the accident flight which results in a prediction of no significant manoeuvre after the Southerly turn near the Northern tip of Sumatra before continuing in a Southerly direction until it ran out of fuel in the Southern Indian Ocean, West of Australia.

The search zone is dependent on the surface area covered by expected descent scenarios from the time of the final satellite log-on attempt at 00:19. This has been defined by expert accident investigators at the ATSB. If the actual descent scenario was inconsistent with the distribution of possibilities considered then the search zone may need to increase in area.

Open Access This chapter is distributed under the terms of the Creative Commons Attribution-NonCommercial 4.0 International License (http://creativecommons.org/licenses/by-nc/4.0/), which permits any noncommercial use, duplication, adaptation, distribution and reproduction in any medium or format, as long as you give appropriate credit to the original author(s) and the source, a link is provided to the Creative Commons license and any changes made are indicated.

The images or other third party material in this chapter are included in the work's Creative Commons license, unless indicated otherwise in the credit line; if such material is not included in the work's Creative Commons license and the respective action is not permitted by statutory regulation, users will need to obtain permission from the license holder to duplicate, adapt or reproduce the material.

© Commonwealth of Australia 2016
S. Davey et al., *Bayesian Methods in the Search for MH370*, SpringerBriefs
in Electrical and Computer Engineering, DOI 10.1007/978-981-10-0379-0_12

References

1. Arulampalam MS, Maskell S, Gordon N, Clapp T (2002) A tutorial on particle filters for online nonlinear/non-Gaussian Bayesian tracking. IEEE Trans Signal Process 50(2):174–188
2. Ashton C, Shuster Bruce A, Colledge G, Dickinson M (2014) The search for MH370. J Navig 60(1):1–22. http://dx.doi.org/10.1017/S037346331400068X
3. ATSB: MH370 Definition of Underwater Search Areas (2014) ATSB Transport Safety Report, External Aviation Investigation AE-2014-054, Australian Transport Safety Bureau
4. ATSB: MH370 flight path analysis update (2014) ATSB Transport Safety Report, External Aviation Investigation AE-2014-054, Australian Transport Safety Bureau
5. ATSB: MH370 search area definition update (2015) ATSB Transport Safety Report, External Aviation Investigation AE-2014-054, Australian Transport Safety Bureau
6. Australian Government Bureau of Meteorology: APS1 upgrade of the ACCESS-G numerical weather prediction system (2012) NMOC Operations Bulletin Number 93, Australian Government Bureau of Meteorology
7. Bar-Shalom Y, Willett PK, Tian X (2011) Tracking and data fusion: a handbook of algorithms. YBS, USA
8. Barber D (2011) Bayesian reasoning and machine learning. Cambridge University Press, Cambridge. http://web4.cs.ucl.ac.uk/staff/D.Barber/pmwiki/pmwiki.php?n=Brml.Online
9. Bernardo JM, Smith AFM (1994) Bayesian theory. Wiley, New York
10. Bertsekas DP, Tsitsiklis JN (2008) Introduction to probability, 2nd edn. Athena Scientific, Belmont
11. Bowditch N (1821) American practical navigator. Government Printing Office, US
12. Bunch P, Godsill S (2013) Particle smoothing algorithms for variable rate models. IEEE Trans Signal Process 61(7):1663–1675
13. Cowan G (1998) Statistical data analysis. Clarendon Press, Oxford
14. Del Moral P (1996) Nonlinear filtering using random particles. Theory Probab Appl 40(4):690–701
15. Doucet A, Godsill S, Andrieu C (2000) On sequential Monte Carlo sampling methods for Bayesian filtering. Stat Comput 10(3):197–208. doi:10.1023/A:1008935410038
16. Godsill S, Vermaak J, Ng W, Li J (2007) Models and algorithms for tracking of maneuvering objects using variable rate particle filters. Proc IEEE 95(5):925–952
17. Griffin D (2015) MH370-drift analysis. http://www.marine.csiro.au/griffin/MH370/
18. Grimmett G, Stirzaker D (2001) Probability and random processes, 3rd edn. Oxford University Press, Oxford
19. Gurley V, Stone L (2015) What does the recovery of floating debris tell us about the location of a wreck?. Memorandum Metron Scientific Solutions, Reston
20. Gustafsson F (2010) Particle filter theory and practice with positioning applications. IEEE Aerosp Electron Syst Mag 25(7):53–82

© Commonwealth of Australia 2016   113
S. Davey et al., *Bayesian Methods in the Search for MH370*, SpringerBriefs
in Electrical and Computer Engineering, DOI 10.1007/978-981-10-0379-0

21. Hu XL, Schon TB, Ljung L (2008) A basic convergence result for particle filtering. IEEE Trans Signal Process 56(4):1337–1348
22. Ihler A (2007) Kernel density estimation toolbox for MATLAB. http://www.ics.uci.edu/ihler/code/kde.html
23. Jazwinski AH (1970) Stochastic processes and filtering theory. Academic Press, New York
24. Jeffreys H (1946) An invariant form for the prior probability in estimation problems. Proc R Soc Lond Ser Math Phys Sci 186(1007):453–461
25. Kalman RE (1960) A new approach to linear filtering and prediction problems. Trans ASME J Basic Eng 82(Series D):35–45
26. Katz J (2010) Introductory fluid mechanics. Cambridge University Press, Cambridge
27. Ljung L (1999) System identification: theory for the user, 2nd edn. Prentice-Hall, Upper Saddle River
28. Lumpkin R, Grodsky SA, Centurioni L, Rio MH, Carton JA, Lee D (2013) Removing spurious low-frequency variability in drifter velocities. J Atmos Ocean Technol 30(2):353–360. doi:10.1175/JTECH-D-12-00139.1
29. Maskell S, Rollason M, Gordon N, Salmond D (2003) Efficient particle filtering for multiple target tracking with application to tracking in structured images. Image Vis Comput 21:931–939
30. Maybeck PS (1979) Stochastic models, estimation, and control, vol 1. Academic Press Inc, New York
31. NOAA (2014) Grid of magnetic field estimated values. http://www.ngdc.noaa.gov/geomag-web/#igrfgrid
32. NOAA (2015) The global drifter program: satellite-tracked surface drifting buoys. http://www.aoml.noaa.gov/phod/dac/index.php
33. Ristic B, Arulampalam MS, Gordon N (2004) Beyond the Kalman filter. Artech House, Boston
34. Roberson B (2007) Fuel conservation strategies: cost index explained. AERO magazine 2:26–28. http://www.boeing.com/commercial/aeromagazine/articles/qtr_2_07/AERO_Q207_article5.pdf
35. Safety Investigation Team for MH370: Factual Information: Safety Investigation for MH370 (2015) Technical Report MH370/01/15, Ministry of Transport, Malaysia. http://mh370.mot.gov.my/download/FactualInformation.pdf
36. Särkkä S (2013) Bayesian filtering and smoothing. Cambridge University Press, Cambridge. http://users.aalto.fi/ssarkka/pub/cup_book_online_20131111.pdf
37. Scharf LL (1991) Statistical signal processing: detection. Estimation and time series analysis, Addison-Wesley, Reading
38. Schön T, Gustafsson F, Nordlund PJ (2005) Marginalized particle filters for mixed linear/nonlinear state-space models. IEEE Trans Signal Process 53(7):2279–2289. doi:10.1109/TSP.2005.849151
39. Shah D (2014) 6.438 algorithms for inference, Fall 2014. Massachusetts Institute of Technology: MIT OpenCourseWare. http://ocw.mit.edu
40. Stone LD, Keller C, Kratzke TL, Strumpfer J (2011) Search analysis for the location of the AF447 underwater wreckage. Technical report Metron Scientific Solutions, Reston
41. Stone LD, Streit RL, Corwin TL, Bell KL (2013) Bayesian multiple target tracking. Artech House, Boston
42. Streit RL (2010) Poisson point processes: imaging, tracking and sensing. Springer, New York. doi:10.1007/978-1-4419-6923-1
43. Tsitsiklis J (2010) 6.041 probabilistic systems analysis and applied probability, Fall 2010. Massachusetts Institute of Technology: MIT OpenCourseWare. http://ocw.mit.edu
44. Van Trees HL (2001) Detection, estimation, and modulation theory. Wiley-Interscience, Hoboken
45. Via Satellite: Inmarsat exec talks about operator role in search for MH370 (2014). http://interactive.satellitetoday.com/inmarsat-exec-talks-about-operators-role-in-search-for-mh370/
46. Vincenty T (1975) Direct and inverse solutions of geodesics on the ellipsoid with application of nested equations. Surv Rev 23(176):88–93

Printed in the United States
By Bookmasters